THE UNITED STATES PULP AND PAPER INDUSTRY: GLOBAL CHALLENGES AND STRATEGIES

by
Jeffrey S. Arpan, *University of South Carolina*
Alan D. Bauerschmidt, *University of South Carolina*
J. Carl Clamp, *University of South Carolina*
Gregory Dess, *Florida State University*
Kate Gillespie, *University of Texas*
Daniel Sullivan, *University of South Carolina*
J. Frederick Truitt, *University of Washington*
and
Eero Artto, *Helsinki School of Business and Economics (Finland)*

A research study and report by the
Center for Industry Policy and Strategy
Division of Research
College of Business Administration
University of South Carolina
Columbia, South Carolina

Funded by

The Business Partnership Foundation
College of Business Administration
University of South Carolina

University of South Carolina Press

First Edition

Published in Columbia, South Carolina, by the
University of South Carolina Press

Manufactured in the United States of America

Library of Congress Cataloging-in-Publication Data

The United States pulp and paper industry.

"A research study and report by the Center for Industry Policy and Strategy, Division of Research, College of Business Administration, University of South Carolina, Columbia, South Carolina. Funded by the Business Partnership Foundation, College of Business Administration, University of South Carolina."

1. Wood-pulp industry—United States. 2. Paper industry—United States. 3. Wood-pulp industry. 4. Paper industry. I. Arpan, Jeffrey S. II. University of South Carolina. Center for Industry Policy and Strategy.

HD9769.W53U638 1986 338.4'76762'0973 86-19296

ISBN 0-87249-501-9

THE CENTER FOR INDUSTRY POLICY AND STRATEGY

In response to increased competition from foreign companies, trade and investment policies by foreign governments, and sometimes even by policies of the United States government, in 1981 the College of Business Administration of the University of South Carolina established a Center for Industry Policy and Strategy. The role of the Center is to conduct global research studies on major industries in order to identify what firms in a particular United States industry can do to improve their international competitiveness and what facilitating policy changes should be considered by the United States government. Thus the focus of each study is both global and industry specific.

Each research study is conducted by a group of at least four faculty members from the University of South Carolina or other universities in the state and at least one foreign scholar/consultant from a country whose industry is a world leader. In addition, an industry Advisory Board is formed for each study to provide additional information and insights to the research team. Thus each study combines perspectives on the future of the industry from several different disciplines (e.g., marketing, finance, management, international business) and viewpoints (academia and business, United States and non-United States). At the end of each study, a national conference is held at the University of South Carolina to discuss the findings and implications of the study, and a book based on the project is published.

The operating premise of the Center is that the international competitiveness of all United States industries can and must be enhanced for the overall well being of the industries and the society. While the studies of the Center cannot in themselves enhance United States competitiveness, they can provide a necessary global overview and analysis which, if properly considered and acted upon by United States firms, can help move the industry in the right direction.

The Center is a nonprofit organization, funded by the State of South Carolina and contributions from the private and government sectors. Contributions from the private and government sectors in terms of time, assistance, and money are both needed and welcomed.

Jeffrey S. Arpan, Director
Center for Industry Policy and Strategy

PULP AND PAPER ADVISORY BOARD

Bobby Abraham*
Director of Corporate Planning
Weyerhaeuser Company

Larry Caldwell*
Former Director of Forest Products Strategy
Continental Group

Gary Caudle
Vice President, International Division
Sonoco

Robert Eisenach
Assistant to the Chairman
Mead Corporation

George Finlay*
Vice President, Operations
Bemis Company

Walt Foody
Manager of Economics
Champion International

Louis F. Kalmar
President
Lou Kalmar Inc.

Manfred Kincaid
Senior Vice President
Albany International

Stewart Monroe
Vice President, International Division
Georgia Pacific

Cliff Morton*
Former Vice President, Corporate Planning and Development
Boise Cascade

Jim Rogers
Vice President, Corporate Planning
James River

Robert Seidl, Consultant
Simpson Timber Company
President (Retired), Simpson Paper Company

*Served as a member of the board for a portion of the project.

PREFACE

Seldom in American history has the United States economy been confronted with so much international competition as it is today. From its oldest industries to its newest ones, American business faces increasingly widespread and sophisticated foreign competition both at home and abroad. Virtually no industry has remained immune to or unaffected by this phenomenon.

The increased internationalization of the US economy brings with it new threats, opportunities, and challenges. As a result, new strategies become necessary at the level of the firm, the industry, and the nation. This study by the Center for Industry Policy and Strategy of the University of South Carolina about the global challenges of the US pulp and paper industry is designed to help increase the awareness and focus the industry's attention on global developments that will continue to affect its competitiveness, viability, and ultimate survival. This international perspective will become increasingly necessary for the industry to understand its strategic alternatives in an increasingly interdependent world. After presenting a global analysis of the pulp and paper industry, several specific policies and strategies are identified by the research team for the US industry and for the governmental agencies charged with the responsibility for enhancing the international competitiveness of US industries.

The study represents the collaborative efforts and perspectives of researchers from the University of South Carolina, Florida State University, the University of Washington, the University of Texas, and the Helsinki School of Business and Economics. This team of researchers was also assisted by a project advisory board comprised of senior corporate executives from the US pulp and paper industry. The conclusions and recommendations of the final report, however, are those of the research team, not the advisory board.

While this final report does represent a group effort and reflects a blending of different viewpoints and contributions, each member of the research team had initial responsibilities for certain aspects of the study:

The US industry: Alan Bauerschmidt, Carl Clamp, Greg Dess, and Fred Truitt.

The Asian Pacific and Canadian industries: Fred Truitt.

The European and Scandinavian industries: Eero Arto and Alan Bauerschmidt.

Government policies and regulations: Kate Gillespie.

International trade patterns: Daniel Sullivan.

Synthesis and project directorship: Jeffrey Arpan.

Two other individuals made significant research contributions to the study: Connie Taylor, a University of Washington Ph.D. student and research assistant to Professor Truitt; and Peter Davis, a University of South Carolina Ph.D. student and research assistant now on the faculty at the University of Oregon. Cheryl Horn also contributed numerous weeks of work in computerizing the industry questionnaire.

The project team is also deeply indebted to the hundreds of pulp and paper executives in the US and abroad who participated in the questionnaire and interview parts of the project, and to the members of the project's advisory board who provided additional information and useful insights. Without their individual and collective participation, the project would not have progressed much beyond being simply an analysis of secondary data.

Finally, the project team is indebted (both literally and figuratively) to the University of South Carolina, its College of Business Administration, and its Business Partnership Foundation for providing the funds necessary to undertake and complete the research project.

Columbia, South Carolina
July 11, 1986

Jeffrey S. Arpan
Project Director

THE UNITED STATES PULP AND PAPER INDUSTRY:
GLOBAL CHALLENGES AND STRATEGIES

Executive Summary

The pulp and paper industries of the world have undergone considerable adjustment over the past four decades as a result of major changes in their competitive environment: the rebuilding of war-torn Europe and Japan, the formation of the European Community (EC), trade liberalization efforts of the OECD nations, environmental protection legislation, energy shocks, differential growth rates in national demand, substitute products, and fluctuating exchange rates. As a result, virtually all countries' industries experienced pressure for structural adjustment, changes in strategy, and political action. The choice and success of adjustment strategies selected varied and were both dictated and limited by national conditions and degrees of international competitiveness. These limitations typically determined whether the adjustment strategies were outward internationally aggressive ones or inward domestically protective ones.

The Canadian industry has experienced a decrease in its comparative advantage and, in all likelihood, will continue to do so in the future. While Canada will continue to be the dominant surplus nation, it faces significant future difficulties in maintaining its export markets, especially in commodity products. Given mixed but preponderantly unfavorable pressures, the Canadian industry's strategic responses were and will probably continue to be more market maintenance than expansion oriented, although aggressive export pricing strategies are probable to keep capacity utilization from falling further in Canada's highly export-dependent industry.

The adjustment pressures and concomitant strategies for Japan's industry were quite different because of its fiber import dependency and surging demand for land, capital, and labor in other, higher growth industries. Under the "administrative guidance" of MITI, a recession cartel was formed to rationalize the industry: reducing absolute capacity, moving toward higher value added products and grades, a host of mergers and acquisitions, and diversification. Its probable future strategic course: penetrate relatively underdeveloped export markets in Asia for lower grade papers; continue to structurally prune, reorganize, and modernize for specialized exports to developed countries; and yield as slowly as possible to foreign competition in fiber-intensive grades.

The export surplus nations of Scandinavia faced foreign government retaliation to their industries and governments' aggressive export strategies of government subsidization, successful industry strategies to increase paper exports more than pulp exports, and successive currency devaluations. These competitive moves put them in even greater competition with the struggling EC industries and led the latter to impose trade protectionist measures. Little change in Scandinavian strategies is likely, given the competitiveness of the industries and the strategic importance of the industries in their nations' economies.

With the exception of West Germany, EC nations' pulp and paper industries continued to lose competitiveness; reeled under import surges from Scandinavia and North America; experienced major declines in employment, firms, and profitability; and sought increased protection. The West German industry's efforts to modernize, prune, and concentrate on market niches lessened its adjustment problems. Little change in any EC country's industry strategy is foreseen in the near future.

Of all the remaining nations in the world, only Brazil and the USSR had any real capability to affect international trade in pulp and paper. Of these two, only Brazil had any major impact as the Soviet Union placed its development efforts in other industries. In fact, the USSR faces the ignominious situation of being a net importer despite having over 60 percent of the world coniferous forests. In contrast Brazil made major strides in investing capital in its industry and related infrastructure and significantly increased its exports. However, an unfavorable investment climate, internal economic problems, and heavy external debt limited its capabilities to become a truly major actor.

As for the US industry, its international competitive position became stronger, both comparatively and absolutely. Continued development of its southern resource base, combined with most other countries' worsening situations, positioned the US industry for an even stronger future. However, while increased capacity is likely to reduce imports to a negligible amount, much of the new capacity will have to be devoted to expanding exports. While the significant comparative advantage of the US industry should make this possible, a continued strong dollar coupled with increased trade protection and industry subsidization abroad could effectively eliminate this potential, causing significant negative repercussions on the US industry and its suppliers. Should this negative scenario occur, the larger US firms would be negatively impacted first. Then, as they attempt to maintain high capacity utilization by more aggressively seeking domestic business, the smaller firms will be adversely affected. Thus smaller firms should seek market niches to enhance their prospects for survival or growth.

As for the larger firms, a continuation of existing strategies is advisable and warranted, with greatest attention needing to be focused on other large US and Canadian competitors' moves. This is especially true for capacity expansions because the major imminent threat/problem is adding too much capacity at once. Other critical strategies necessary involve the industry's relationship with the US government:

1. Obtaining governmental pressure to reduce existing import barriers abroad and/or keep them from rising further.
2. Monitoring "unfair" trade practices of competitor nations' industries in terms of their export markets to the US and to countries where the US industry is also exporting.
3. Keeping US capital, energy, and regulatory compliance costs at a competitive level with those of other major countries.
4. Keeping the foreign exchange value of the dollar from rising further, which, if unchecked, would seriously impact US export potential.

Given the existing and increasing international competitive advantage of the US industry and that much of the future expansion of US industry capacity will have to be export market focused, maintaining and enhancing the US industry's well-being will require more US government support internationally in the areas enumerated above.

TABLE OF CONTENTS

CHAPTER 1

INTRODUCTION

During the past two decades, most American industries have been subjected to increasingly successful foreign competition at home and abroad. While some American industries have successfully adapted to this change in their competitive environment, many other industries have not, and virtually no industry has been immune to, or unaffected by, foreign competition. Many export markets for US products have declined while imports from abroad have reached record levels, and the profitability of most US firms and their foreign affiliates has been negatively impacted. In short, the international competitiveness of American industry has declined, with generally negative repercussions on the United States.

Many reasons are offered to explain the deterioration. They include "unfair" competitive prices initiated and supported by foreign governments; declines in the quality of US products, productivity, and technological innovativeness; a lack of international orientation and awareness among the ranks of US management; a bias by US management for short-term profits and goals at the expense of long-term objectives; an overvalued dollar; and inconsistencies in the domestic and foreign policies of the US government.

Regardless of the single or combined accuracies of these and other explanations, what appears necessary to improve the international competitiveness of some US industries is a restructuring of the way things should be done. More specifically, most US industries need to acquire an international perspective of their individual industries, their competitors, and the dynamics of competition.

In an era of rapid economic, political, and technological changes, and with the increase in uncertainty that these changes bring about, it is a formidable challenge to anticipate problems, identify key issues, and arrive at solutions that will enhance the international competitiveness of particular industries. This report is the product of the effort of the Center for Industry Policy and Strategy to meet this challenge for the US pulp and paper industry. It presents an analysis of the major international changes that are occurring in the economic and political forces that directly affect competition in this industry and concludes with a set of policy and strategy alternatives designed to enhance the US industry's competitiveness at home and abroad.

THE PURPOSE OF THE STUDY

The primary purpose of this research was to gain a better understanding of the global competitive dynamics of the pulp and paper industry and its competitive structure in order to formulate future policy and strategy alternatives for companies in the industry. In general, the study was designed to identify and analyze (1) the critical changes occurring in the forces that affect the global and domestic competitive environments of the industry; (2) the policies and strategies that US companies, major foreign competitors, and their governments are using or developing; (3) the policies and strategies that are likely to be successful in the future; and (4) the strategic choices US companies should consider adopting if US international competitiveness is to be enhanced. Government policy initiatives supportive of the strategic choices recommended are also identified and discussed.

COUNTRIES INCLUDED IN THE STUDY

While data on the global pulp and paper industry were collected, the US, European, and Japanese pulp and paper industries received the major focus. The European and Japanese industries were selected because of their international importance _vis-a-vis_ the other pulp and paper industries of the world and because they have had to make major adjustments in recent years. The adjustments that were made and the political options considered and finally implemented by their governments provide the US industry with a set of tested policy and strategy scenarios. The actual outcomes of these scenarios, although strongly influenced by local economic, social, and political factors, provide information on what policies and strategies have and have not been successful, and why.

MAJOR RESEARCH QUESTIONS

If the US pulp and paper industry is to enhance its international competitiveness in the future, it must be able to respond flexibly to a wide range of domestic and international political and economic changes. Central to this study was the identification and analysis of those changes critical to the competitiveness of the US industry and its flexibility, and the question of how to respond to the favorable and unfavorable effects of changes.

The issues addressed in this study were of two distinct types: (1) those fundamental forces that, from a practical perspective, cannot be altered but need to be taken into account in any policy planning, and (2) those constituting problem areas that can be addressed amd resolved, at least in part, through proper decisions and which need timely policy attention.

Against a backdrop of rapid fundamental changes in the domestic and international competitive environments of the industry, there are many problem areas requiring resolution. What political, cultural, social, and economic factors adversely affect the US industry's development and response? What policy changes could be made by the US government to remove competitive obstacles and to enhance the industry's ability to respond? What constitutes practical responses by US pulp and paper producers?

Specifically, the major research questions addressed in this study included the following:

1. What fundamental changes occurred to the global pulp and paper industry during the last two decades that directly and/or indirectly affected the competitiveness of the US industry?

2. What factors currently restrict and/or impede the US pulp and paper industry's ability to respond to the fundamental changes that are occurring?
3. What policy role could the US government play in removing current obstacles in order to enhance the competitiveness of the US industry?
4. What effective policy and strategy adjustments can and should be undertaken by US pulp and paper producers?

RESEARCH APPROACH

The study represents the cooperative efforts and perspectives of a research team consisting of seven faculty members and four graduate research assistants. To ensure that the research and the resulting policy and strategy recommendations were of pragmatic value to the US pulp and paper industry, the research team was advised and assisted by an Advisory Board consisting of 13 senior operating executives from 12 of the US industry's leading public and private corporations.

The research consisted of three parts. One part examined the trends in structural factors and market forces that determine the intensity and dynamics of the competition in each national industry. In general, these structural factors and market forces are the same in all national industries but vary in their levels and degrees of importance. Competition in an industry is the result of the dynamic interplay of five basic competitive forces: (1) the relative bargaining power of the industry and its customers, (2) the relative bargaining power of the industry and its suppliers, (3) the threat of new entrants, (4) the threat of substitute products or services, and (5) the level of rivalry among existing firms. Further, the strength of these forces, individually and combined, is determined by the structural features of the industry and includes such things as entry and exit barriers, economies of centralization of functional activities, capital requirements, switching costs (from one industrial sector to another, from one product line to another), access to distribution channels, the pervasive and fundamental influence of production and product technologies, government policies, rules and regulations, and so forth.

The second part examined the differences in selected factors and forces of national industries that determine the intensity and dynamics of international competition within the global pulp and paper industry. These differences included: (1) factor cost differences among countries (such as wage rates, costs of other inputs, capital-labor combinations, and productivity); (2) differing circumstances in the national markets such as supplier relations, market sizes, and sociocultural factors; (3) different roles that national governments play (such as incentives, subsidies, and degree of protection); (4) differences in the rates at which new production and product technologies are introduced; and (5) differences in the goals, strategies, and resources of national and international competitors, and their ability to monitor and react to foreign competitors and changes in the environment.

The first two parts of the study depended heavily on the analysis of secondary data and information and used such traditional techniques as trend analysis, comparative analysis, and the interpretative abilities of the researchers and practitioners.

The third part of the study examined the domestic and international strategies of major competitors in the global pulp and paper industry. The varying abilities of the competitors to adapt to, take advantage of, or otherwise influence competition within their industry were of critical importance. Eventual success is directly linked to the understanding competitors have of their industry's market forces and structural factors, the changes that are occurring in them, and the strategic alternatives they are prepared to consider and select from in order to compete.

The third part of the study relied heavily on the analysis of data and information collected from interviews with executive officers of North American and European pulp and paper producers and from questionnaire responses received from more than 300 US producers. In order to understand the process by which firms react to their environments and competitive conditions, it is necessary to examine actual industry perceptions. Hence the need for primary data collection. In addition to the interviews and questionnaires, other information was obtained from companies' annual reports, presentations by their officers, and related articles appearing in the mass media.

In selecting the companies to be interviewed, an effort was made to ensure that they were representatives of the types of pulp and paper producers to be found in North America and Europe. The main criterion was, however, a willingness to cooperate and frankly discuss variables related to strategy in the industry. Clearly, the sample of companies is not representative in any statistically rigorous sense. The findings and conclusions derived from a study of corporate strategies must be interpreted within the broader environmental conditions prevailing in the global industry, in light of other studies and the experience of the readership.

More detailed information is provided in subsequent chapters about the specific research methodologies employed. In addition, a statistical supplement containing the responses from more than 300 US firms to the research questionnaire is available separately from the Division of Research, College of Business Administration, University of South Carolina.

ORGANIZATION OF THE REPORT

Chapters 2 through 6 present the findings, conclusions, and recommendations arrived at as a result of this study. An analysis of world production, consumption, and trade patterns dating from the 1960's is presented in Chapter 2, followed by a study of regional patterns. Notwithstanding the influence of trade barriers on trade patterns, such an analysis provides insights into the relative and changing competitiveness of pulp and paper industries. It also provides a necessary backdrop to the more detailed analyses presented in subsequent chapters.

Chapter 3 deals with the economic factors and conditions which affect the shape and character of national pulp and paper industries and their potential degree of international competitiveness. The discussion and analysis move sequentially from the global to the regional, and then to the major countries within each region.

Chapter 4 examines the influence that national government policies have on the international competitiveness of their pulp and paper industries, whether these policies are domestic or international in nature. It identifies policies which help and those which hurt the pulp and paper producers.

The results of the corporate interviews and questionnaire surveys are presented in Chapter 5. The presentation includes a composite of executive perceptions of the strengths and problems in the pulp and paper industry currently and in the future, opportunities and risks in the industry, and possible actions the industry may take with respect to its problems.

Chapter 6 draws together the findings and conclusions presented in the previous chapters in order to identify the most logical strategy alternatives for the major pulp and paper industries of the world and concludes this study by presenting a series of recommendations for the US industry. These recommendations reflect the cultural, social, economic, and political setting of the US and our interpretations of global economic, technological, and political trends. Modifications in the recommendations are also considered under different economic scenarios. The types of government policies which would help to support these recommendations are also identified.

CHAPTER 2

INTERNATIONAL TRADE AND THE PULP AND PAPER INDUSTRY

A phenomenon affecting most of the world's industries since the mid-1940's has been increased international competition. To be sure, some industries have been more acutely affected than others, but the days of national economic self-sufficiency are clearly over. Virtually no nation's industries are immune to international shocks, problems, and opportunities despite numerous and varied attempts by national industries and governments to the contrary. The change effected by the progressive internationalization of any industry has many implications for the efforts of producers to improve, let alone preserve, their market competitiveness. Narrow strategic focus on positioning the firm within the national market against national companies is a management approach with fading relevance. The rising volume of sales in the home market contested for by an increasing number of foreign companies strains the conventional notions of competing. A new set of concerns demands acknowledgment. For instance, at the macro level of the firm, offensive concerns (deciding on and managing the possible involvement of the firm in the international market), as well as defensive concerns (responding to invasion of the home market by foreign competitors or even local companies' forays abroad), now demand a more significant consideration in managerial decisions. This is because the type of decision reached regarding international market opportunities and foreign competitive threats shapes the competitive character of the enterprise. Thus a comprehensive study of any nation's industry must be cast in terms of global conditions, and this study of the US pulp and paper industry is no exception. It is simultaneously a study of the US and global pulp and paper industries.

This chapter provides an international background by examining past and current trends/patterns in the international trade of pulp and paper. It begins with an overview of aggregate trade patterns, followed by trends in major product categories, then the trade patterns of leading regions.

The chapter concludes with an overview of the major forces underlying the international trade of forest products, pulp and paper. These forces are discussed in greater detail in subsequent chapters.

THE SIGNIFICANCE OF INTERNATIONAL TRADE

An increasingly important agent of change in the global pulp and paper industry is international trade. Contrary to the beliefs of some industry participants, international trade cannot be relegated to second-level priority considering that in 1981 20 percent of the total worldwide production of forest products entered the international market, an increase of 25 percent from the level registered 25 years earlier (see Exhibit 2-1). With an estimated market value in excess of $50 billion, the sales and the profit performance of all but the most protected firms were affected (see Exhibit 2-2). Indeed, as the president of Copaced (the French Paper Industry Confederation) observed, regarding the rise of domestic sales captured by foreign producers, "the industry is open to the world, irreversibly."[1] Statistics and sentiment attest that the penetration of national markets by foreign competitors is a *fait accompli*; the task at hand is to assess the impact and its future implications for the industry.

The Product Composition of International Trade

Indiscriminate usage of the generic terms "pulp," "paper," and "board" is misleading in discussing international trade in forest products. The varieties of pulp, paper, and board produced and the demand for each among nations give international trade a distinct character that is impossible to pinpoint without analyzing the specific product classes traded. Identifying the products traded is all the more pressing because space limitations require aggregate measures of pulp and paper volume and value as the principal units of analysis to be used throughout this chapter. Therefore, awareness of the dominant characteristics of foreign trade of forest products is prerequisite to the usefulness of the discussion of the foreign trade profiles of regions and leading nations.

The prominence, insignificance, and emergence of major product classes for pulp, paper, and board on the international market are outlined in Exhibits 2-3 and 2-4. Statistics indicate that international trade in pulp occurs across six classes. But two classes, chemical pulp and wastepaper, compose the bulk, with the former accounting for most of the pulp trade. In 1971 and 1981 trade in any combination of chemical grade pulp represented two-thirds of the total trade in pulp, be it sulphate, sulphite, or soda; bleached, semi-bleached, or unbleached; coniferous or nonconiferous. The dominance of chemical pulp is due to the product's high value to weight ratio relative to that of competing mechanical or semi-chemical pulp grades. The greater surplus of profit over transportation costs diminishes the constraining influence of this barrier to international trade. In addition, the numerous grades of paper and board whose production requires highly processed grades of chemical pulp prompt the large volume entering the international market. The statistics of Exhibit 2-3 show that trade in mechanical pulp continues to account for a smaller proportion of international trade. In the early 1950's this pulp category held a 15 percent share; 25 years later its share had slumped to under 5 percent. The slide in mechanical pulp will continue as the primary consumer group, newsprint producers, uses advanced forestation techniques to cultivate an ample domestic supply of trees which are processed into mechanical pulp and nearby mills at a much lower cost.

The most striking development in pulp trade is the emergence of wastepaper as a trade commodity. In 1981 this low-cost alternative to virgin wood chips and pulp constituted 20 percent of international pulp flows. Furthermore, the demand for wastepaper appears to be gathering momentum as evidenced by its increasing share of total trade. Between 1971 and 1981 wastepaper was the only pulp category to gain market share. The recent importance of wastepaper is attributable to pollution control measures and conservation attitudes in many nations consuming high volumes of paper. Also, improved recycling processes which increase the fiber yield and reduce processing costs add an attractive economic incentive. A case in point is the US. The US is the greatest consumer of paper products in the world and generates an enormous quantity of wastepaper.

In 1981 exports of this commodity surpassed two million tons representing 58 percent of the total pulp exports and having a transaction value in excess of $180 million. Thus, the environmental and economic rewards of pulp recovered from wastepaper have made this the foremost type of pulp export from a nation whose virgin pulp supply is virtually inexhaustible. In due course wastepaper will emerge as the single dominant pulp commodity traded in the international market, particularly as nations such as Japan and the UK push wastepaper recovery rates toward 50 percent and beyond.

The corresponding pattern of international trade in paper and board is seen in Exhibit 2-5. These statistics show that the combined share of three product groups--newsprint, printing and writing paper, and packaging paper and board--represents more than 90 percent of the paper and board market internationally. Although their combined share has remained stable over the period surveyed, their relative proportions have not. Newsprint in 1971 accounted for 44.6 percent of paper trade, but by 1981, this share had fallen 20 percent. The declining share of this commodity is initially puzzling in light of the growth in worldwide newsprint production capacity. Over the same ten-year period, the two leading producers of newsprint, Canada and the US, registered respectively a 14 percent and 51 percent expansion in newsprint production volume. One recent development appears to be responsible for the continuing decline of newsprint as an international trade commodity: The growth in the trade of wastepaper and the improved quality of recycled stock have created a comparatively cheap production input in the manufacture of newsprint. Therefore, for many nations, the cost of producing their own newsprint has become competitive with importing newsprint from leading producer countries. The lower cost of raw materials and the improved manufacturing capability of many nations formerly dependent on foreign-made newsprint have lessened international trade in this commodity.

The cultural paper class (especially printing and writing papers) has captured the lost market share of newsprint. Between 1971 and 1981 the 87 percent rise in the absolute volume traded pushed this category's relative share of the total paper trade to 23.2 percent. Moreover, the change in relative share increased 35 percent from 1971. This paper class is made up of numerous products with thousands of uses, ranging from the special paper used in printing national currency to commodity lines such as light coated paper used in magazines. The prevailing demand in all national markets for printing and writing papers, coupled with the differing character and comparative advantage of each nation's paper production capability, designates this an important product class in international trade.

Consumers in many nations have given low priority to converted products, and this is responsible for the low trade share of the remaining paper product categories, notably household and sanitary papers. However, the improved standards of living in the majority of these markets promise to fuel the demand for specialized forest products. The predominant focus of the forest production industry in most of these nations on producing industrial paper and board indicates that satisfying domestic demand will force recourse to foreign supply sources.

The trade in wrapping and packaging board dwarfs the foreign trade activity in all other board categories. Similar to the trade pattern in cultural paper, trade in this industrial product class is not restricted to one major product type. Cross-national shipments take place across many grades with the leading two, folding boxboard and kraftliner, each accounting for roughly a quarter of total board trade. The remaining flow is generated by a half dozen board grades, each targeted toward more focused market niches. The two remaining industrial paper and board product classes, construction and other, are insignificant to aggregate international trade. Generally, the demand for these two specialized products is met by the national paper industry or through substitutes from the actively traded wrapping and packaging paper and board grades.

THE TRADE PATTERN OF LEADING REGIONS

Statistics do show consistent, albeit slow, growth in the amount of forest products entering international trade both as a relative and absolute proportion of aggregate output. Environmental forces, particularly the irregular boom-bust economic cycle, have affected the magnitude of this flow, but they never have reduced international trade to insignificance. However, these forces have precipitated notable changes in the trade pattern of leading regions in forest products. Exhibits 2-4 and 2-5 outline the advances and declines in the relative importance of each region. The measure for the importance of a region is its share of the total dollar volume of forest products imported and exported in five-year intervals between 1966 and 1981. The slated future status of each region is drawn from its forecasted balance of trade for pulp and paper for 1985 and 1990 (see Exhibits 2-6 and 2-7). Statistics on 1978 performance are included for comparison. Data in the body of these two tables represent the past and projected sum of exports less imports for the corresponding regions. In addition to sketching the general pulp and paper trade patterns for major regions, the unit of measurement, tonnage, in Exhibits 2-6 and 2-7 contributes a needed measure of scale.

Import Pattern of Leading Regions

Some regional markets have had supply shortages because of their small indigenous forest or their limited ability to process available timber stock. These markets have created a demand pull which has been instrumental in expanding the world trade of forest products. When an unfavorable supply-demand disequilibrium emerges in a regional market, importing from foreign markets becomes more imperative. The sophistication of economic and social activity in the various nations comprising the region is an excellent indicator of when this threshold is crossed: the more advanced their overall development status, the greater the demand for cultural and industrial paper and board products. This positive relationship between national development and domestic consumption underlies the historic pattern of regional import activity.

The largest importing region has been Western Europe. In spite of intraregional production increases of 87 percent and 125 percent for pulp and paper respectively between 1960 and 1982, Western Europe is caught in the throes of worsening import dependence: The rapid growth in consumer and industrial needs has pushed demand beyond the maximum feasible yield of the indigenous forest production base. Western Europe is more dependent on imported forest products than any other region. In 1981, the sum of import transactions with fellow bloc members and the world showed that Western Europe was involved in over half the forest products traded on the international market. Asia ranked second, but its share was well under half that of Western Europe.

The trade deficit of Western Europe is characterized by product imbalance. That is, forest products imports are conspicuous only in one major class: pulp. The importation of fiber is so significant that dependence on foreign supply sources is the norm for the majority of nations within this region. Paper and board trade activity, however, follows a different path. The numerous production facilities throughout the region have perennially generated amounts of finished products in excess of intraregional demand. Success in this area has limited the damage done to the regional balance of trade. Nevertheless, it has not prevented the emergence of an asymmetrical trade profile. The question then arises: Why the asymmetry?

The historic strategy of Western European manufacturers directs industrial activity toward capitalizing on the chief resources of the region: production expertise, capital, and market demand. In exploiting that which is abundant, many European producers have achieved distinctive competence in making finished paper and board goods from imported pulp. Their major problem is the lack of raw material available within the region. Hence the need to import pulp. Forecasts call for the trade profile of Western Europe to become more exaggerated as dependence on pulp soars and intraregional demand soaks up more finished goods, thereby leaving less to export. In 1978, for instance, the total forest product deficit of Western Europe was only 91,000 tons; by 1990 this is expected to surpass 2.5 million tons, an increase of 2600 percent. Paper exports will not offset growing pulp imports. Barring the development of a synthetic substitute for pulp, Western Europe will suffer a worsening trade deficit in forest products, a fate that will further designate Western Europe the most import-dependent region in the world.

In direct contrast to the predicament of Europe, the involvement of North America in forest product imports is diminishing. The statistics in Exhibits 2-7 and 2-8 reveal a trend toward progressive import independence. National demand, in this case that of the US and Canada, is met with limited recourse to the international market. In 1966 the production shortfall across select pulp and paper grades compelled this region to claim 22.4 percent of the forest products imported that year, but a steady rise in production capacity had, by 1981, driven down this share to only 15 percent. Declining import dependence is attributable to three characteristics of this region: first, the abundant endowment of timber resources; second, the improving yields of forest tracts brought about by improving forestation technique; and finally, the deep pool of investment capital available to finance the operation of the largest forest production complex in the world. Achievement in these three areas has allowed North America to minimize dependence on foreign market sourcing.

Apparent in the North American forest product industry's investment activity is a drive to go far beyond self-sufficiency in forest products. In light of current and planned expansion, the era of import independence for North America is dawning. For instance, the negligible import volume of pulp will be so dwarfed by the export side that North America is predicted to have a nine million ton pulp surplus by 1990, and the anticipated surplus of paper and board will exceed eight and a half million tons. Thus, the effort of North American producers to establish and expand productive capacity is forcing a decline in the import share of this preeminent regional market. The industry's performance suggests that eventually the future involvement of this region in the import dimension of international trade will be insignificant.

The multiplicity of factors which affect the consumption of forest products and their differing intensity of interaction in each region has changed the relative import share of regions. For the large part, changes have been minor, accompanied by limited repositioning in the cardinal ranking of regions. One change, however, has been startling: the dramatic increase in the significance of the Asian market in the import segment of foreign trade activity. Its mediocre share of 14 percent in 1966 increased 80 percent by 1981. Asian markets are now involved in over 21 percent of total import transactions in forest products. The only change of comparable magnitude over the same period was the 32 percent drop in the import share of North America. Apparently, the emerging Asian region is capturing the surplus made available by the declining import share of North America. Undoubtedly, the rising participation of Asia in the international market will continue. Therefore a greater proportion of the world's forest product output is destined to cross Asian borders.

The dependence of Asia on other regions' surplus pulp is due to the chronic shortage of easily accessible timber. Although within the Asian region much fiber remains to be tapped (notably in the forest areas of Indonesia, Australia, and New Zealand), wide stretches of forest lands remain inaccessible because of climatic and infrastructure barriers. Hot, muggy locales without adequate industrial or personal support facilities make exploitation a costly proposition. Also hampering the efforts of interested parties is the reluctance of host governments to improve conditions. Because pulp mills are among the most expensive ways of creating new jobs, national leaders, strapped for capital, give pulp mills low priority. Together the social, capital, and governmental barriers have delayed starting promising projects.

Asian paper and board making firms have established trade linkages with nonregional pulp producers in response to the worsening shortage. These sources are generally on the opposite side of the Pacific Rim: Japan in 1981 was the leading customer of Brazilian pulp mills; China likewise with Chile. Also, Asia claims much of the pulp surplus of North America. But the result of closer trade ties with foreign suppliers is greater dependence on pulp, which by 1990 is expected to manifest itself in a regional pulp deficit surpassing four million tons.

The outlook for paper and board imports also is severe in spite of the achievements of the Asian forest product industry. Over the past two decades, leading consumer nations have committed to processing imported pulp and developed an extensive converted product manufacturing capacity.

Intraregional demand, however, always has surpassed on-line capacity. For the immediate future, the addition of new plants will not narrow this unfavorable demand-supply gap. Thus the endless chase continues because as living standards improve throughout the region, they are stimulating a previously dormant demand for converted products. Items such as newspapers, magazines, soft-tissue, sanitary products, writing papers, and so forth, are being sought actively by more of the consuming population.

The momentum of Asian market demand is gaining strength from the newly industrializing countries in the region and also from the volume of foreign investment flooding Asian nations. In 1982 the flow of capital from the US to Asia surpassed (for the first time ever) the US investment flow to Western Europe. The general acknowledgment that the traditional East to West flow has been reversed permanently is projected to fuel the primed Asian economies. Rising prosperity will increase greatly the demand for finished paper goods. The lack of sufficient production capacity, present and planned, to fulfill inevitable needs is expected, by 1990, to balloon the regional paper trade deficit up 170 percent over the deficit level recorded in 1978. Given this and the previously discussed pulp deficiency, it is clear that the Asian market represents the most vital regional import market as the world heads into the twenty-first century.

Export Pattern of Leading Regions

To round out the regional picture, this section discusses the past and future export behavior of major regions in forest products. Of course, an analysis of the degree of involvement of certain regions in the import of pulp and paper underscores profound developments in the pattern of international trade. Further, it designates promising commercial targets for the marketing efforts of firms intent on exploiting the opportunities in international business. Narrow concentration on the import dimension, however, blinds analysis to an important consideration: regions whose resource character and industrial capability position them to be effective competitors for sales. Moreover, sales competition is not restricted to third-nation markets, as in the case of North and South

American companies vying to capture long-term supply contracts for Japan's fiber requirements. Of more pressing urgency is gauging the likelihood of penetration by foreign producers of a firm's home turf. For instance, South American producers' heated pursuit of sales opportunities in North America increased South American exports of forest products to North America 44 percent between 1976 and 1981.

The statistics in Exhibit 2-5 concerning regional export of forest products reveal that the dominant proportion of world trade is supplied by the two foremost producing regions--North America and Western Europe. The preeminence of these two regions has changed little over the past generation. In 1966, for example, their combined share of aggregate global exports (that is, the amount of forest products shipped across national frontiers on an intra- and interregional basis) totaled 78.1 percent; by 1981 their combined share had grown to 82 percent. Their direct involvement in export transaction volume surpassed 40 million tons of forest products with an estimated market value exceeding $39 billion. Disguised by the immense scale of their combined share, however, are fundamental features that distinguish the export pattern of North America from that of Western Europe.

North American producers are capturing a larger share of world sales of forest products because of the tremendous production surplus in forest products. Improving the chance of sales success abroad is the competitive edge North American producers hold over most pulp and paper firms in other countries. North American producers have comparative and in many cases absolute cost advantage in manufacturing a range of products. The economies of scale resulting from the immense size of an increasingly integrated industry coupled with the abundant supply of low-cost production inputs make North American products among the most cost competitive in the international market. Protectionist trade barriers of other nations do lessen this advantage, but numerous products from North America continue to offer an attractive alternative in virtually all overseas markets. Therefore, only in the unlikely event that the forest lands of North America are decimated by disease or drought will North

American producers not stake out a greater export share in the global pulp and paper market.

Western Europe is confronted by a different set of conditions which threaten to precipitate a decline in the export competitiveness of Western Europe forest products in the international market. The principal source of trouble arises from the previously discussed salient dependence of Western Europe on foreign pulp sources. The cost distortion created by this handicap restricts the foreign market scope of many regional producers. The expense of long-distance transportation raises the cost of pulp. This in turn results in higher priced output which compromises the competitiveness of numerous converted products outside the tariff-protected national or regional--that is, the EC--markets.

The inflated cost-tariff structure relationship restricts the converted product exports of most West European nations to those countries that are cosignators on custom duty reduction agreements. The high price of many West European products relative to alternatives in foreign markets makes interregional export prospects dim. As a result, Western Europe faces a bleak export future.

South America steadily is transforming its tropical jungles into a managed forest production base yielding an immense quantity of low-cost raw materials. Progressive exploitation of domestic resources slowly is turning this region into an important force in the international trade of forest products. Signs of success are evident already. Between 1976 and 1981 South American exports of pulp products to leading developed nations increased 270 percent; paper and board exports increased 104 percent. South America accounted for a larger share of total world exports because of the marked increase in regional export volume. These accomplishments, however, were offset partially by increased imports of softwood pulp products (up 33 percent) and paper and board (up 31 percent) over the same five-year span.

The improving exports performance of South America is threatened by the general economic difficulties rampant throughout the region. The burden of a huge foreign debt has halted expansion programs in all markets. For example, in 1982 Brazilian producers, citing the lack of low-interest long-term financing, halted or cancelled all major new investment. Current difficulties, however, alter neither the resource endowment of the region nor the commitment of government and industrial representatives to capitalize on it. Indeed, it is inevitable that South American producers will carve out a larger niche in the global forest products market.

A puzzling trend revealed by statistics in Exhibits 2-6, 2-7, and 2-8 is the sputtering export performance of the African continent. Exporting nothing in 1954, by 1966 African producers had captured a 3.4 percent share of aggregate world exports. A decade later, however, there was a 20 percent drop in share, declining even further by 1981. These statistics become more portentous when the performance of South Africa is factored out. This nation gradually has increased its capacity and, in 1981, by itself generated 64 percent of the continent's total pulp, paper, and board output. Moreover, of the remaining 36 percent, the North African subregion exploited its close proximity to needy European markets and accounted for roughly 25 percent. Therefore, the vast bulk of the continent which lies in the fertile tropic zone produced only 10 percent of continental forest product exports, of which pulp predominated.

The low per capita consumption of paper and board throughout Black Africa, along with the general political turbulence, has created market conditions unlikely to reverse the decline of Africa. Economic and political underdevelopment hamper the emergence of integrated, cost-effective pulp mills and paper plants necessary to make products that are price competitive on the world market. Still, as with Latin America, the promise of the region is seductive: The combination of favorable climates and the preponderance of rich soils close to coastal harbors contributes to the excellent potential of the region. The major problem lies in transforming potential into reality, a process with which Black Africa will long struggle. Meanwhile nominal exports of pulp, almost exclusively from North Africa to Europe, will dominate the export profile of Africa.

Trade Profile of Prominent Nations

Understanding the global flow of forest products requires an appraisal of the foreign trade activity of the major regional markets. Such macro analysis, however, is limited in pinpointing underlying trade linkages. This concern is especially important for the pulp and paper sector given the large volume of intraregional as opposed to interregional trade. Therefore, it is important to supplement regional analysis with focused study of the foreign trade activity of the individual nation. For the sake of expediency, only those prominent nations whose past performance and present character indicate that they will condition the future pattern of international trade in forest products will be evaluated.

United States

US producers have a leading trade position in several forest product commodity markets. From the chemical pulp family, to paper for publications, to paperboard products, the US presence in the international market is acute. In absolute quantity the magnitude of trade is remarkable; the volume of the annual trade flows of the US surpasses the aggregate production and consumption of all but a few nations. The level of American exports and imports, however, pales in comparison to the massive production and consumption volume of the US market. In 1981, for instance, forest product imports represented just 10 percent of consumption and exports accounted for under 7 percent of production (see Exhibits 2-9 and 2-10).

Theoretically, the US, if pressed, could follow an autarkic trade policy, electing to isolate its market from the world and, through domestic production, suffer no supply shortfall. In 1981, the net trade deficit in forest products was just over four million tons. Meanwhile the operating rate of pulp and paper mills hovered at about 90 percent in producing 103 million tons. Thus, plants would need only to raise operating rates 4 percent to cover the net imported forest products. However, instead of autarky, US producers have followed the logic of classical trade theory, seeking to exploit the gainful opportunities of international trade. This has resulted in an asymmetrical pattern of trade with respect to nations that purchase from and supply the US.

The import profile of the US is straightforward and simple because of domination by a single trade partner: Canada. Perhaps in no other American industry sector is the trade linkage with the northern neighbor of the US so encompassing. Shipments from Canada account for over 90 percent of the total pulp and paper imports of the US. Imports of pulp and paper from all other nations are insignificant; following Canada is Finland with less than one tenth of 1 percent share of the aggregate imports of the US.

Trade reciprocity with Canada is impossible because domestic consumer demand is covered two to three times by national producers. Consequently, the export activity of US firms has followed a pattern different from that of imports. The largest export markets for US pulp producers have been the leading developed countries in pulp-starved Western Europe and Asia. In 1982 these two regions together accounted for 70 percent of US exports, but due to regional differences in economic growth rates, the export importance of each is changing. The proportion of American pulp destined for the four leading West European consuming markets--the UK, Italy, France, and the Federal Republic of Germany (FRG)--has fallen from a 51 percent share in 1966 to 37 percent in 1981. Offsetting this slide has been the booming demand of Asian markets, the destination for only 27 percent of US pulp exports in 1976, but 33 percent in 1981.

The emerging export pattern of US pulp producers--declining share West, increasing share East--also characterizes the export of paper. However this reversal has been more dramatic. The five-year span of 1976-1981 saw US exports of paper to Asia rise 125 percent in absolute volume. More impressive was the 200 percent rise in the significance of the Asian market; from 1976 to 1981 the share of US paper exports shipped to Asian markets jumped from 12 percent to 36 percent. Conversely, the export significance of the big four European consumer nations declined.

In 1976, the four accounted for 22 percent of US paper exports, and by 1981 their share had dropped to 19 percent.

The final export region of significance to US producers is South America. Although in absolute terms exports to South America increased 150,000 tons between 1976 and 1981, this region's relative share of US aggregate exports fell from a 29 percent share in 1976 to 27 percent in 1981. The development of indigenous production facilities is improving the self-sufficiency of the South American region. A consequence already apparent is the declining dependence of South American consumers on US producers for finished and converted forest products. This trend will continue unless, of course, substantial cross-regional price differentials emerge. This latter development is a remote possibility given the strong commitment among South American governments and industry leaders to stake out a major presence in international trade of forest products.

Whether these trade trends will continue is questionable in light of the regional realignment under way in the US forest products industry, specifically, the shift of American companies' control and production locus to the southern half of the US. More than ever US firms are reducing their reliance on the slow-growing coniferous forests of the Northwest, Northeast, and upper Midwest and are depending more on the hospitable South and bountiful Southern timber. The primary stimulus of this shift (62 percent of new production capacity is slated for the South) is the tremendous growth potential of this region in terms of trees and, therefore, profits. The abundant sunshine of the sunbelt provides a 12-month growing season, four more months than in the North. Sunbelt trees thereby grow to cuttable size in 20 rather than 40 years. Further bolstering productivity is the development of "supertrees": quick-growing, disease-resistant trees that yield great amounts of pulp. By 1990 every seedling planted is slated to be one.

The conducive effects of the environment further enhanced by scientific forestation techniques have convinced Edward A. Gee, Chairman of International Paper, "that by the year 2010 the [Southern] plantations will be producing twice as much wood per year as they do today."[2] Given the inherent leverage capability of pulp (that is, a small amount of fiber is needed relative to the weight of the finished product), the implication of bountiful wood becomes even more startling. The ascendancy of the South has prompted American producers to build region-based world class facilities merging computer technology with advanced presses and dryers. Future output is projected to be both prolific and produced at the lowest cost in the world.

It is doubtful that the demand in the American market will match production increases; forest product companies will have a large surplus of trees or pulp or paper. To avoid inventory buildup or, worse, low capacity utilization, American producers will have to pursue sales in foreign markets, especially in those developed countries outside the tropic zone. Several nations with temperate climates already operate with forest product deficiencies and will not be able to increase the yield from indigenous forest lands to meet projected market demand. In other words, there are prime candidates, notably in Western Europe and Asia, to whom surplus production may be sold.

The proximity of large tracts of Southern forests to port facilities on the Eastern Seaboard and the Gulf of Mexico will expedite the delivery of product from plant to port. Significant cost saving in this activity is assured. Harvesting and transportation costs in the Northwest, for instance, average three to five times those in the South. The reduced processing expense of Southern wood products on top of gains expected through resource and industrial productivity will sharpen the price competitiveness of this output on the international market.

The shift to the sunbelt will be decisive in reshaping the character of the trade profile of the US, but to what degree is hazardous to speculate. On the North American continent, one consequence, however, is already apparent in the trade relationship between the US and Canada.

Canada

Throughout the 20th century the forest products industry of Canada has depended on the expanding economy of the US to provide broad-based market demand. Hitching onto the US economic engine has enabled the forest products sector of Canada to grow and prosper more quickly than prevailing national conditions would have warranted. The southward migration of the American forest product industy has caused Canadian producers to refocus their traditional marketing strategy.

Export data for Canada reveal an effort to move away from dependence on the economy of the US by targeting other national markets for forest products sales. The drive to diversification is evident in the decreasing proportion of Canadian exports delivered to the US market. In 1966 Canada shipped 89 percent of total pulp exports and 77 percent of total paper and board exports south to the US. By 1981 export dependence on the US had fallen to 77 percent for pulp and 52 percent for paper. Interestingly, Canada's forest products imports, generally a negligible amount of specialty products purchased exclusively from the US (over 99 percent), are dropping as the export significance of the US market declines. From 1976 to 1981 Canada's imports of US paper fell 40 percent following the 86 percent rise in the previous five-year span.

Offsetting slackening US demand for Canadian forest products are markets in Asia, South America, and Western Europe, the exports to which have increased 75, 28, and 22 percent respectively from 1976 to 1981. Therefore, to underscore an important development, a rising volume and proportion of Canadian forest products are being diverted from historic intraregional marketing channels. Instead, they rapidly are entering the international marketplace. The volume of this diverted flow--nearly three million tons in 1981--is substantial enough to alter conditions in the international market.

Hastening the diversification effort of Canadian producers is their development of specialized ocean-going bulk vessels that make overseas transportation economical. These vessels are designed with the sole purpose of transporting forest products and have altered favorably the cost of shipping forest products from West and East Coast provinces to export markets abroad relative to the shipment of goods to the US by rail. An improved, cost-effective delivery capability will strengthen the competitiveness of the overseas presence of Canada and, thereby, will impel transformation of the trade profile of this leading pulp and paper producer nation.

Japan

A review of the data detailing the relative significance of imports and exports to the national market as presented in Exhibits 2-6 and 2-7 indicates that the international market is incidental to Japanese forest product manufacturers. In 1981, foreign trade in pulp and paper reached just 16 percent of aggregate production. On a relative standing, then, the involvement of Japan, like the US, in the foreign trade of forest product is a secondary preoccupation involving marginal tonnage. However, when the absolute volume of this marginal tonnage is noted, it is apparent Japan's foreign trade activity, again like that of the US, is a key determinant of the supply and demand equilibrium in the international forest product market. But Japan has implemented an export program differing from that of the US. Consistent with the well publicized national industrial policy of their nation, Japanese forest product companies have concentrated on developing an export base in high-value converted wood goods. This strategy has shown remarkable success. In 1971 under 6,000 tons of paper products were earmarked for export, but by 1981 export volume had increased 125 times, up to 654,000 tons. Concomitantly, imports of converted products registered only a fivefold increase, even though national demand rose much more. Becoming a volume exporter of high value-added products in spite of indigenous raw material scarcity has forced Japan to accept spiraling trade deficits in pulp. The inability of national producers to generate sufficient quantities of production inputs and the abundance of Japan's three main suppliers--US, Canada, and Brazil--assure that pulp deficits are a permanent feature of the trade profile of Japan.

Scandinavia

The bloc of nations popularly called "Nordic Suppliers Inc."--Sweden, Finland, and Norway--significantly influences the nature of pulp and paper trade in Western Europe. Their low aggregate paper consumption, coupled with the vital need to exploit the prime indigenous forest bases, has made the Nordic industry thoroughly export oriented. The combined total exports of the Scandinavians accounted for close to 80 percent of pulp exports and over 50 percent of paper exports within the European region in 1981. Clearly the competitive threats manifest in intraregional nations' drive toward self-sufficiency in finished products and the prevalence of lower priced pulp from distant sources have not usurped the commanding regional market position held by the Scandinavians. However, there are developments under way in trendsetting Sweden and Finland which foreshadow a profound alteration in this market position. In light of the prominence of these two countries these changes hold dire implications for the competitive tension in and, more fundamentally, the character of forest products trade in Western Europe.

Because of large surpluses in production, Sweden has built up a tremendous flow of forest products to Western Europe and, to a lesser degree, abroad. The character of this flow, however, is in transition because of a heightened national awareness that virgin fiber resources were being utilized to their productive limit. The concern of government and industry representatives over potential long-run damage to the balance of trade and also to the prized forest production base has forced revision of the historic export strategy of Sweden. Readily exportable bulk pulp has become less important, and highest priority has been given to increasing exports of more profitable finished forest products. To hasten this transition Sweden drove down foreign demand for pulp through the infamous "scissors pricing" strategy. In brief, Swedish producers raised the foreign price of pulp proportionately more than that for paper and board. The relative price inelasticity of pulp notwithstanding, pulp exports declined. The growing stockpiles of pulp then were diverted to domestic paper producers whose operating strategy was geared explicitly for export of high-profit finished products.

Trade data confirm the consequence of this realignment in export priorities; between 1976 and 1981, the export of pulp from Sweden tumbled 15 percent versus the 6 percent rise recorded from 1971 to 1976.

Concomitant to the drop in pulp exports between 1976 and 1981, export of paper and board soared 46 percent compared to the 17 percent rise registered in the preceding five-year period. Sweden's drive toward increasing the export of paper, coupled with the almost universal integration of their forest production complex, confirms that this new trade pattern--rising paper export, declining pulp export--is not a temporary aberration. Rather it presages the long-term export profile of the Swedish forest products industry.

Corresponding data for Finland initially suggest that Finnish pulp and paper producers are ignoring the Swedish lead and, instead, are electing to uphold the tradition of Nordic Supply Inc., that is, increasing both the pulp and paper stock available to consumers in foreign markets. Between 1976 and 1981 export of pulp soared 63 percent while export of paper rose 35 percent. Proponents of the high-value product export strategy point out that this is a dangerous policy, but Finnish authorities appear to be cognizant of the inherent threat. As a former Minister of Trade and Industry admitted, there was only one realistic way Finland was going to maintain a healthy balance of trade and that was to increase the value added to exports. The question then arises: why the divergence between policy and practice?

Finnish producers are implementing a broader interpretation of the trade strategy of restricting exports to high-value forest product classes. Forest producers in Finland have applied the logic to both paper and pulp, the latter as confirmed in the product mix of pulp exports. Specifically, in 1981 mechanical pulp, a low-profit pulp grade, constituted 40 percent of the total pulp produced in Finland. Mechanical pulp, however, accounted for less than 1 percent of aggregate pulp exports. In contrast, lucrative chemical pulp, 60 percent of national pulp production, constituted over 95 percent of pulp exports.

The long-term feasibility of this policy is doubtful. Currently the flood of low-price pulp into the region from North and South America is putting downward pressure on profit margins. Finnish authorities' concern over damage to their balance of trade, along with the depressed profits the national industry is garnering, will compel national producers to follow the course of organized and efficient industries and switch valuable production inputs into the highest value processes. Therefore, the trade profiles of Sweden and Finland will gradually converge, a development promising to alter fundamentally the nature of forest product trade in Western Europe.

Already their partial convergence has triggered profound change. Consider that in 1966 the combined pulp exports of Finland and Sweden totaled 5.13 million tons, surpassing easily that year's paper exports of 3.9 million tons. But by 1981, this ratio had been reversed dramatically: Pulp export had sunk to four million tons while export of paper had soared 125 percent, up to 8.5 million tons.

The release of 1984 trade data will reveal an acceleration of this trend because on January 1, 1984, a major trade barrier which had hampered paper exports of the Scandinavians to their principal export market was removed: The EC repealed tariff restrictions levied on newsprint imported from Nordic nations. Unrestricted access gives Nordic newsprint producers (all three of whom are in the top five producer nations of the world) a significant cost advantage over their chief competitor, Canada. In 1982, Canadian producers shipped 721,000 tons of newsprint to Western Europe, which earned them a 20 percent market share versus 35, 26, and 15 percent shares for Finland, Sweden, and Norway respectively. If matching legislation granting Canada the same tariff exemption privilege is not forthcoming from the European Community, the European market position of Canadian producers will be vulnerable. Current market conditions indicate the eventual outcome will be Western Europe awash in Nordic produced newsprint.

Western Europe

Counterbalancing the perennial surplus of the Nordic bloc is the permanent supply shortfall experienced in neighboring continental and offshore nations. Generally, these countries[3] are plagued by forest production bases incapable of generating enough raw material to satisfy national demand for finished products. Under conditions prevalent in lesser developed countries notorious for low per capita consumption of paper, most European nations have ample timber stock to meet reduced national needs. The prevailing level of economic activity and social development, however, taxes the indigenous forests of each of these countries far beyond their maximum yield. To overcome this lack of timber, these nations have implemented, to varying degrees of success, a common foreign trade strategy. Its dominant traits and principal implications will be more easily seen by looking at the record of the two nations whose performance records offer the most startling contrast, the UK and the FRG.

The United Kingdom. With the exception of Ireland and the Netherlands, the UK has a smaller proportion of forest land than any other European nation. This scarcity of land is the principal cause of the pronounced dependence of the UK on foreign trade for nearly 90 percent of the pulp and 60 percent of the paper consumed domestically in 1981. A minor degree of independence has been achieved through extensive wastepaper recycling; in 1981, this effort generated 2.2 million tons of raw materials, more than eight times the total virgin pulp milled. Notwithstanding this contribution, the reliance of the UK on foreign suppliers--notably Canada, the US, and Sweden--is extreme.

A recent development indicates that this situation will worsen. The struggle of the British pulp and paper industry against inefficiency has resulted in numerous plant and mill closures. Therefore, in the one area of possible competence, paper and board making, the performance of the British industry has been dismal; between 1970 and 1980, the output of finished goods from UK plants declined 23 percent while the corresponding output of France, the FRG, and Italy increased 25, 36, and 43 percent respectively. Naturally, the proportion of paper and board supplied by imports jumped 51 percent during the decade. One somewhat favorable consequence of declining plant capacity was a 28 percent improvement in the pulp trade deficit from 1970 to 1980. When measured in pounds sterling, however, the improvement has damaged the balance of trade by forcing the import of expensive finished products to replace lost domestic production.

The trend toward declining pulp imports and rising paper and board imports will dominate the future foreign trade activity of Britain because UK producers, intent on escaping the competitive handicap of scarce high-priced pulp and also the persistent difficulties in the national economy, are scaling down investment in UK-based capacity. Instead, they prefer to target investment in more attractive overseas locales. If the flight of precious capital continues, the UK will become permanently dependent on foreign-produced finished wood products. This is significant for it would make the UK the single most important import market for paper and board.

The Federal Republic of Germany. The foreign trade activity of the FRG follows a path similar to yet distinct from that of the UK. This nation, like many of its continental neighbors, has overcome the disadvantage of a small indigenous forest through distinctive competence in making paper and board. Concerted effort among industrialists to structure an industry around its primary resource had made the FRG so self-sufficient in select areas that surplus is available for export. Still, in other areas imports are the principal source. Data in Exhibits 2-9 and 2-10 show the rough balance between national production and imports. In 1981 imports net of exports as a percentage of aggregate forest products production was 40 percent for the FRG; the UK had a 133 percent share. Thus, West German-based production exceeded net imports by 60 percent, whereas in the UK net imports exceeded the total amount of forest products processed.

The activity of the FRG in the international market is driven by two motivations: in pulp, economic need; in paper and board, political convenience. Desperate for pulp, the FRG imports this commodity from a cross-section of suppliers in the Nordic areas as well as in North and South America. On the other hand, trade in paper and board is conditioned by the prospective trade partners' relationship with the EC. The reduced or repealed custom duties on trade of goods with nations affiliated with the Community is decisive in promoting exclusive trade with economic allies. The price competitiveness of a West German-made product in the French or Italian markets vanishes when a 5 or 10 percent tariff surcharge is slapped on at delivery to a nonaffiliated market. Given this, it is not surprising that almost 90 percent of the paper and board trade of FRG producers occurs within the confines of the EC.

Brazil

One nation on the other side of the globe, Brazil, gradually is becoming a forcible actor in the international trade of foreign products. Until recently its prime national resources--vast tracts of quick-growing tropical forests made accessible via the seaward flowing Amazon--had been exploited only marginally. Restraining the effort was the lack of developmental capital to finance construction of the industrial capacity needed to transform raw material into marketable products. Beginning in the 1970's an influx of capital in the form of well-publicized loans from foreign banks, and supplemented by substantial foreign direct investment, contributed to the critical startup financing. Soon capacity expansion was racing and production volume began to inch upward. The Brazilian economy's concurrent rapid expansion fueled domestic market demand for forest products. Almost all forest products made went to satisfy national needs, thereby reducing the quantity for export.

Current economic conditions are not quite so healthy and Brazil, with immense capacity on line, has turned toward offloading surplus production into the international market. Moreover, the need to generate earnings in foreign currencies and to improve the sick balance of trade makes the increasing of exports even more important. These conditions have made Brazilian companies aggressive international marketers, chasing sales opportunities around the globe. By 1982, 46, 25, and 16 percent of Brazilian foreign pulp sales were made to Western Europe, Asia, and North America respectively. Although industrial growth has slowed, expanding on-line plant capacity and progressive sophistication in forest management have positioned Brazil to capture a greater proportion of pulp and paper sales in the world market.

Union of Soviet Socialist Republics

To round out the country analysis section, it is necessary to evaluate the status of a potentially dominant force in the global pulp and paper market, namely, the USSR. Potential must be stressed in light of the historically marginal performance of this nation. Although endowed with roughly 60 percent of the coniferous forest of the world, the Soviet Union produces barely 6 percent of the world volume of forest products. The track record of the USSR in foreign trade also is grim. In 1966 the proportion of world imports of forest products to the USSR was under 1 percent; by 1981 its share of world imports had risen to 1.7 percent. Export performance is worsening also; the Soviets' share of total global export sales dropped over 20 percent from 1966 to 1981. Finally, the USSR is projected to be a net importer of pulp and paper by 1990. Their 1981 performance in pulp created a sizable surplus, but in paper they quickly are approaching deficit status as evidenced by the slim 4.7 percent surplus recorded in 1982.

To avoid the humiliation of the most timber-rich country becoming a net importer, the USSR has begun an ambitious program to boost processing capacity. Following the outline in their current five-year economic plan (1981-1985), efforts are under way to raise pulp capacity by 35 percent, from 7.1 to 9.5 million tons; paper capacity by 25 percent, from 5.2 to 6.5 million tons; and board capacity by 44 percent, from 3.4 to 5 million tons.[4] Complete success is doubtful, however, given the profound weaknesses in supporting infrastructure and the historic failure of the Soviets to achieve stated goals. Production inadequacies compounded by the stark urgency of fulfilling growing domestic demand for consumer-oriented products will constrain USSR activity in foreign trade. Therefore, the Soviet Union, despite a superior resource base and close geographic proximity, poses no credible competitive threat to North American and European producers battling for market share in the booming Asian region.

FORCES PROMOTING INTERNATIONAL TRADE OF FOREST PRODUCTS

The chief presumption of analysis throughout this chapter is that the intensity of a country's involvement in international trade is a direct function of one central relationship: the product yield of the national forest production base and the level of aggregate national demand. If supply falls short of demand, then market forces pressure for imports; if supply exceeds demand, market forces pressure for exports. Implicit in this simplified abstraction of a complex market process is that all nations search for balance, search for equilibrium between production supply and consumer demand in the domestic market. International trade provides the vehicle to attain balance by facilitating factor movements between nations who have production surplus and their counterparts who suffer a shortfall in supply. The presumption that international trade is driven by this single force is supported continuously by the foreign trade behavior of regions and nations. To end the explanation here is an injustice to the diversity and complexity encountered by forest products companies active in the international market. Four other forces, each conditioned by the central relationship of the resource endowment and the level of market demand, also must be evaluated to underscore more precisely the market motivations which propel the progressive internationalization of the pulp and paper industry. These four forces--consumer preferences, synthetic resource barriers, the "lean times" phenomenon, and cross-national cost differentials--are reviewed next.

Consumer Preferences

The preferences of consumers for wood commodities unavailable from domestic sources is a leading stimulus of international trade of forest products. Unavailability, which can be partial or total, arises from two conditions: (1) the precise input requirements of production and (2) product quality standards. Precision in raw materials translates into international trade when the production of consumer-demanded goods depends on inputs available exclusively from foreign sources of supply. For instance, in certain grades of industrial packaging products--wrapping, bag, linerboard, corrugated medium--neither strength nor durability can be sacrificed. This necessitates using production inputs with the specific character of that from the long-fibered softwood tree species cultivated in nations with temperate climates. Localized supply but worldwide demand results in the anomaly of tropical nations endowed with extensive forest regions actively trading in the international market for pulp. In 1981, for example, pulp imports into short-fiber pulp rich South America reached 500,000 tons. The market of origin for 99 percent of this flow was three temperate zone nations: Canada, the US, and Norway.

The second dimension of consumer preferences promoting international trade is product quality standards. The purchasing behavior of consumers continually demonstrates that producers must meet product quality standards transmitted by market demand. Fulfilling this task often forces companies into the international market to source the needed product. The trade pattern of the US offers striking testimony: American forest products companies' annual production of hardwood lumber so far exceeds national demand that the US ranks as the leading exporter of this wood grade in the world. Nevertheless, the US ranks consistently as the leading national importer of hardwood lumber. This paradox occurs because consumer demand outstrips the domestic supply of certain grades of hardwood lumber. In spite of the abundant availability of hardwood lumber grade as a substitute production input, the product quality expectations of consumers made substitution a suboptimal production strategy. Therefore, the pursuit of domestic sales prompts forest product manufacturers to source needed hardwood lumber grades from foreign markets who have a surplus.

Synthetic Resource Barriers

Meeting national demand without involuntary recourse to foreign trade depends on the level of natural factor endowment exceeding domestic need. However, this alone is insufficient to give a country relative import independence. Three other conditions are also decisive: capital for technology acquisition and financing operations, a competent managerial class, and a sophisticated national infrastructure system.

When each of these four conditions--the first by fate and the latter three by effort--cross some threshold, only then can the industrial capability to transform natural resource into usable products develop.

Mirroring the unequal national endowment of forest lands, synthetic resources also are not distributed equally among all countries. Indeed, only a select few nations operate with no handicap in any of the synthetic resource classes; the vast majority of countries are self-sufficient in some but not all. Limitation in any class, however, establishes internal market barriers that make necessary involvement in the international trade of forest products. The problem of deficiency in just one synthetic resource is best exemplified by the plight of the Soviet Union. The USSR has one quarter of the world's forests and records annually the greatest woodcutting volume of all countries. But in 1981, the Soviet Union had only a slight trade surplus in forest products; moreover, it is projected soon to attain deficit status. Poor performance is attributable to the inability of Soviet companies to access great stretches of the inner-continental northern forest region on a cost-efficient basis. The economic feasibility of industry expansion into this desolate Canadian-sized area is compromised by the enormous cost of developing even a rudimentary infrastructure system. At many isolated sites the Soviets are confronted with desolation and must build roads, bridges, power stations, and entire townships to make a profitable go of it. It has been rumored that, because they are desperate to avoid this immense expense, the Soviets have commissioned studies to evaluate the probable cost and environmental impact of reversing the flow of the Volga river. The northward flow of the Volga limits its usefulness for transporting felled timber to the existing complex of pulp and paper plants based in the south. Despite the contribution of less ambitious programs, infrastructure problems haunt the Soviet Union's forest production efforts to the degree indicated by their growing dependency on the international market.

In many nations the limitation extends across all three major classes of synthetic resources. The lack of capital funds and of managerial and infrastructure sophistication forces vast tracts of indigenous forest lands to remain untapped, despite the escalation of national need. This predicament is particularly persistent throughout several wood-rich nations, notably those located in the tropic zone encircling the globe. Complicating matters is the general political and market instability in many of these nations; hazardous national business environments give rise to levels of risk that daunt even the most daring financiers. Investment in plants and people is, therefore, slow to materialize. Consequently, countries having abundant natural resources but little developmental capital face the perverse problem in which import of forest products is more economically feasible than domestic production. International trade, therefore, becomes the only avenue to gain needed forest product for domestic consumption.

The "Lean Times" Phenomenon

The forest product industries of the US and, to a lesser extent, Japan have been geared traditionally to meeting the voluminous demands of their respective markets. The character of their industrial complexes also is a powerful force periodically driving upward the volume of international trade. Producers in the US and Japan have invested much capital to develop huge capital-intensive facilities capable of significant scale economies. In 1981 the absolute level of their combined production capacity for paper surpassed 80 million tons while that of pulp was near 65 million tons. Naturally, the US and Japan consistently claim the top two positions in the annual ranking of the most prolific producer nations. During periods of strong domestic demand this capacity is invaluable as plant utilization rates approach their upper bounds. Moreover, bulging order files and tight supply inspire confidence among some firms to implement capacity expansion programs.

Trouble arises when the level of national economic activity begins trending downward. The inherent cyclical dynamic of the modern economic system periodically drives market demand below levels that are able to absorb producers' supply of pulp and paper. Faltering demand unleashes the bane of all capital-intensive industries, overcapacity. Aggravating this problem is the completion of expansion projects begun in better days and which come on line as demand is slackening. This unfortunate occurrence struck the US in 1980, when capacity increased 3 percent even though national consumption of paper and pulp was weakening, eventually falling 2 percent by the year's end. Rising overcapacity further depresses plant utilization rates toward levels making it difficult for companies to sustain favorable price structures. If downtime forces price slashing, it is often impossible for companies to gain the profits deemed necessary both for reinvestment purposes and to provide a reasonable return on capital. Admittedly, this is a gloomily drawn scenario; but to the occasional misfortune of the forest products companies, it is realistic.

The strong relationship between the economic well being of pulp and paper producers and the constancy of capacity utilization above the break-even point is decisive in periodically elevating the importance of international trade. The effort to minimize downtime promotes earmarking for export a greater proportion of output heretofore marketed domestically. In the mid-1970's Japan employed this strategy and was able partially to offset the damage of domestic hard times through frenzied export activity. In the midst of the economic recession of the mid-1970's triggered by the OPEC oil price hike, paper production of Japan dropped over 15 percent in one year. At the same time exports of paper more than doubled in volume. Furthermore, the share of exports relative to total production registered a 200 percent increase. Upon the resurgence of the national economy exports dropped but still remained at higher levels than those preceding the oil crisis. US producers also have displayed international trade behavior consistent with the lean times phenomenon. Their repeated demonstration prompted one industry commentator to note, "Traditionally slow US demand is inevitably expressed by the presence in Europe of spot quantities"[5] as US firms rush to offload surpluses overseas.

A minor point of clarification is necessary: Even though the additional tonnage that the US or Japan ships overseas constitutes a small proportion of their respective output, the absolute quantity is sufficiently large to alter supply-demand conditions in the world market.

For example, if in 1981 the US had been 10 percent long on paper, then 5.5 million tons of output could have been dumped into the international market, a quantity greater than the combined paper exports of the FRG, the UK, Italy, France, the Netherlands, and Spain.

Recent market developments indicate a diffusion of the lean times phenomenon to another prominent forest product manufacturer--Brazil. The severe financial problems afflicting Brazil have restricted the national economy and precipitated a drop in consumer demand for forest products. Government and industry representatives' desperation to avoid an idle forest production complex led them to exploit foreign market demand, particularly in Western Europe. Their efforts have resulted in solid penetration of hardwood pulp in Europe at the expense of indigenous softwood grades. Brazilian processed hardwood pulp is estimated to have accounted for more than half of the market pulp consumed in Europe in 1983.

To expect that Brazilian producers will forsake this commanding market share upon the resurgence of the home economy is foolish. Therefore, as seen in the case of Japan, periodic lean times in select nations is a vital stimulus promoting expansion in the international trade of forest products.

National Cost Differentials

Classic international trade doctrine contends that when the price of a good produced domestically exceeds the price of a comparable foreign-made product plus transportation costs, the basis exists for gainful international trade. The purchasing behavior of consumers around the globe demonstrates continuously that when the opportunity for gain is available it is exploited. Although the apparent spark of foreign trade activity is cross-national price differentials, the underlying proximate cause is absolute differences in cost across national markets. This is particularly so in the pulp and paper industry, in which the overriding determinant of the market price of output is cost of production because, unlike some industries (notably automobiles and electronics), the pulp and paper industry must predicate its price structure on the cost incurred in production and distribution. Competition is too keen to do otherwise.

The powerful effect of cost differentials on international trade is due to the disparity in the product cost structure across producer nations. The presence of disparities is crucial, for if insignificant, considerable pressure for international trade would dissipate. In terms of the commodity product kraft paper grade market pulp, salient differences distinguish the cost position of four preeminent producer nations (see Exhibit 2-20). Within this exclusive group, in which each member is endowed with an enviable forest production base, it is the interplay between market conditions and synthetic resources that results in the differences of 1 to 35 percent characterizing production cost of this pulp grade. Excluding Sweden, the three other nations surveyed in Exhibit 2-20 represent the world's lowest-cost producers of kraft paper grade market pulp. This achievement has had positive consequences for the competitiveness of their forest products in foreign markets.

Consider the case of US producers who are fiercely competitive in making cost-effective output. Although unable to claim absolute advantage in making this pulp grade (Canada and Brazil registered slightly lower cost for softwood and hardwoods respectively), their comparative cost advantage with regard to the rest of the world further boosted their absolute cost advantage in manufacturing kraft liner. A favorable cost structure in processing the key production input gave US paper and board companies the most price-competitive kraft liner in the international market. In 1982, the international market share of US-produced kraft liner was almost double that of the only other major competitor, Sweden. All other nations, operating with comparative cost disadvantages, were unable to compete for market share and, as a result, held minor or insignificant positions.

From the perspective of consumers, capitalizing on cross-national cost differentials represents a major way to get the biggest bang for the buck. The favorable cost structures of American companies producing bleached softwood kraft recently offered European consumers an attractive product option. In the first half of 1982 the listed price of US output of this product was $362 per ton versus the price of $462 and $483 per ton for Finland and Sweden respectively. At the risk of belaboring the obvious, net of transport expense European consumers stood to save upwards of 10 to 20 percent by purchasing US rather than regionally produced bleached softwood kraft. Their adherence to economic logic was instrumental to the US producers' strong international sales performance in this product grade. In summary, differing production cost structure for producer nations, a permanent situation due to the unequal distribution of natural and synthetic resources, provides an irresistible economic incentive that drives both producers and consumers to seize the gains available through international trade.

In theory, cross-national differentials in pulp and paper are attributable solely to the variance in cost structures of the national forest products industries. In reality, the explanatory power of theory is limited by two factors: the intervention of governments into market activity and fluctuations in the value of a nation's currency relative to that of its trading partner. The trade policy adopted by political officials concerning the flow of factors across national frontiers often determines the composition, direction, and magnitude of international trade. Whether the policy focus is to promote the development of the national industry by subsidies of financial and administrative assistance or to protect the national industry from foreign competition through tariff and nontariff barriers, the consequence is the same: An artificial cost system is instituted which subverts the basis for international trade created by cross-national cost differentials.

A company backed by public funds is not bound to act in a pattern consistent with standard laws of competition. Instead of producing up to the breakeven point (where marginal cost equals marginal revenue), for instance, a subsidized company is free to produce at points below marginal cost and thereby cut prices while running an economic loss. A subsidized company therefore bases pricing decisions not on its production cost structure but rather on the amount of aid forthcoming from political officials. Although the notion of governmental subsidy is opposed strongly in the US and the FRG, in many nations subsidies, either direct or disguised, are standard operating procedure.

Trade policy, like subsidies, pursues the same outcome but relies on different techniques. In 1983, for instance, European forest products companies charged that American kraft liner manufacturers were dumping their surplus into the European market. In retaliation they lobbied successfully the EC to mandate a minimum price level below which no kraft liner, regardless of original source, could be sold throughout the Community. Following passage of the tariff petition, foreign-produced kraft liner was subject to a higher tariff upon import into any nation affiliated with the EC. The slapping on of a price surcharge--the sale prices of US producers had been below the new price minimum--forced onto American suppliers an artificial competitive handicap that offset partially their comparative cost advantage earned in the production of kraft liner.

Also affecting cross-national cost differentials are the fluctuations in the value of a nation's currency relative to that of its trading partners. Exchange rate change is a key determinant of the price competitiveness of products sold in foreign markets. Forest products producers are threatened by two sources of exchange rate risk, one natural, the other artificial. Natural risk refers to exchange rate change precipitated by market conditions such as cross-national differences in the rate of inflation and interest. For instance, relatively low levels of inflation coupled with high rates of interest in one nation drive up demand for that nation's currency and thereby the exchange value of the currency. The effect of high interest-low inflation on currency fluctuations is best illustrated by the current plight of US pulp and paper producers. With inflation under 5 percent and real interest rates almost double that, the dollar has been performing strongly. The unwavering appreciation of the dollar over the past three years, however, has saddled a 25 to 35 percent price premium on American forest products purchased in foreign markets. The emergence of the superdollar consequently has driven down US producers' export of pulp and paper from the $4.9 billion peak recorded in 1981 to $4.2 billion in

1983. Furthermore, a decline in export value to below $4 billion is forecasted for 1984. Therefore, regardless of the superior cost position of US producers, disfavorable currency changes have lessened, and in some cases offset, their comparative advantage in foreign markets.

Artificial exchange risk, on the other hand, alludes to the effort of a national government to alter unilaterally the relative value of its currency through subversion of market forces. A growing trade deficit, political heat from imperiled export-oriented industries, and high unemployment, for example, pressure the government to improve the export position of national industry by altering the value of the national currency. Specifically, instead of allowing market forces to determine the exchange value of the national currency, governments occasionally resort to "monetary dumping"--large and unjustified currency devaluation openly targeted at improving the export competitiveness of national industry. The inevitable outrage of trade partners aside, monetary dumping strengthens the export position of domestic producers operating with a comparatively high production cost structure.

Recently, regarding forest products, the Nordic nations have been the most flagrant users of monetary dumping. In October 1982 the Swedish government devalued the krona 16 percent. The Finns, claiming the need to maintain currency alignment with their major trade partner, followed suit shortly thereafter with a 10 percent devaluation. Government intervention thus gave a tremendous boost to the declining export competitiveness of the politically salient and economically vital forest products industry.

To gain perspective on exchange rate fluctuations a cautionary note is warranted. Changes in exchange rates have both favorable and unfavorable impact; while the export position of one nation strengthens, those of its trade partners weaken. But as the recent past has dramatically underscored, relative currency strength or weakness is a fleeting condition. Only five years ago the great decline in the value of the dollar had the international market in an uproar; now the reemergence of the superdollar, once again, has the international market in an uproar. Therefore the effects of exchange rate change on cross-national cost differentials can be profound but are not permanent. Indeed, the adage "what goes around, comes around" describes aptly the nature of fluctuations in foreign exchange values.

Footnotes

1. _Pulp and Paper International_, Volume I, No. 25.

2. "The Paper Industry on a Roll," _The New York Times_, December 4, 1983.

3. West European countries referred to are Austria, Belgium, Federal Republic of Germany, Ireland, Netherlands, Italy, United Kingdom, France, and Luxembourg.

4. _Pulp and Paper International_, Annual Review: Europe, 1983.

5. _Pulp and Paper International_, September 1981.

EXHIBIT 2-1

The Significance of International Trade
to the Forest Products Industry
(as measured by the proportion of world output
entering into international trade)

Year	Pulp (%)	Paper and Board (%)
1956	15.3	17.4
1961	15.9	16.9
1966	17.4	19.8
1971	15.2	20.7
1976	15.1	16.0
1980	18.0	19.6
1981	20.4	19.8

Source: International Trade Yearbook, United Nations, various issues.

EXHIBIT 2-2

Value of World Imports and Exports of Forest Products
(Thousands of U.S. Dollars)

Year	Imports	Exports
1966	$ 9,657,455	$ 8,555,620
1967	9,925,330	8,825,531
1968	11,228,248	9,956,719
1969	12,718,838	11,296,826
1970	14,146,381	12,562,059
1971	14,755,343	13,062,760
1972	16,958,850	15,444,306
1973	24,938,337	22,436,465
1974	32,754,583	29,137,884
1975	28,745,270	25,868,648
1976	34,535,680	31,278,058
1977	38,705,808	33,882,763
1978	43,095,567	37,917,067
1979	55,149,803	49,256,144
1980	61,961,847	55,682,198
1981	55,252,470	51,232,034

Source: FAO Yearbook of Forest Products, United Nations, various issues.

EXHIBIT 2-3

Volume of Pulp Entering International Trade by Category
(000 tons)

Pulp and Wastepaper	1971	%	1981	%
Mechanical pulp	962.3	7.5	1063.9	4.9
Semi-chemical and chemi-groundwood	104.7	1.0	120.1	0.6
Chemical pulp	8628.5	67.9	14538.3	67.3
Dissolving pulp	1581.6	12.4	1381.5	6.3
Other pulp	68.3	0.7	132.1	0.7
Wastepaper	1343.8	10.5	4365.6	20.2
Total	12689.2	100.0	21601.5	100.0

Source: *The Pulp and Paper Industry*, Organization for Economic Cooperation and Development, 1972 and 1982.

EXHIBIT 2-4

Volume of Paper and Paperboard Entering International Trade by Category
(000 tons)

Paper and Paperboard	1971	%	1981	%
Newsprint	9767.9	44.6	10405.0	35.2
Printing and writing papers	3646.2	16.6	6851.4	23.3
Construction paper and board	82.9	0.5	186.4	0.6
Household and sanitary paper	288.1	1.3	384.3	1.3
Thin papers	96.7	0.5	216.6	0.7
Wrapping and packaging paper and board	6816.4	31.2	10282.0	34.7
Other paper and paperboard	1166.4	5.3	1238.2	4.2
Total	21864.6	100.0	29563.9	100.0

Source: *The Pulp and Paper Industry*, Organization for Economic Cooperation and Development, 1972 and 1982.

EXHIBIT 2-5

The Import Share of Major Regions in the International Trade Flow of Forest Products

Regions	1966	1971	1976	1981
North America	22.3	19.5	16.4	18.0
South America	2.0	1.9	1.7	2.1
Europe	52.7	50.8	53.0	50.0
Africa	3.0	2.9	3.8	3.5
Asia	14.0	18.9	20.4	22.9
U.S.S.R.	1.0	1.3	1.5	1.7
Oceania	1.8	1.7	1.3	1.0

Source: FAO Yearbook of Forest Products, United Nations, 1966, 1971, 1976, and 1981.

EXHIBIT 2-6

The Export Share of Major Regions in the International Trade Flow of Forest Products

Regions	1966	1971	1976	1981
North America	34.1	33.9	32.8	33.2
South America	2.7	2.3	2.5	2.9
Europe	44.0	42.6	43.0	43.2
Africa	3.4	3.6	2.7	2.2
Asia	8.0	11.0	12.0	11.6
U.S.S.R.	7.2	6.6	6.0	5.0
Oceania	0.5	1.0	1.0	1.5

Source: FAO Yearbook of Forest Products, United Nations, 1966, 1971, 1976, and 1981.

EXHIBIT 2-7

Net Regional Trade in Pulp and Waste Paper 1978-1990
(1,000 tons)

Region	1978	1985	1990
North America	+6078	+7440	+9000
South America	-536	+610	+1815
Western Europe	-2764	-3625	-4720
Eastern Europe	-415	-640	-780
Africa	+85	+123	+135
Asia	-1656	-2995	-4065
Oceania	-562	-825	-1030

Note: A plus sign indicates that the region was and will be a net exporter, while a minus sign denotes a net importing region.

Sources: Compiled from various data sources.

EXHIBIT 2-8

Net Regional Trade in Paper and Board 1978-1990
(1,000 tons)

Region	1978	1985	1990
North America	+4358	+7379	+8731
South America	-1891	-2281	-2091
Western Europe	+2673	+2436	+2181
Eastern Europe	-151	-655	-647
Africa	-1088	-1083	-1108
Asia	-2586	-4738	-7187
Oceania	-353	-27	+771

Note: A plus sign indicates that the region was and will be a net exporter, while a minus sign denotes a net importing region.

Sources: Compiled from various data sources.

EXHIBIT 2-9

Imports, Exports, and Trade Balance of Pulp for Selected Countries, 1981

	Consumption (000 tons)	Imports (000 tons)	Imports as a Percentage of Consumption	Production (000 tons)	Exports (000 tons)	Exports as a Percentage of Production	Trade Balance (000 tons)
Canada	12,661	141	1.1	19,275	6,591	34.1	+6,450
U.S.	47,329	3,707	7.8	46,979	3,336	7.0	-371
France	3,253	1,651	50.7	1,699	154	0.09	-1,497
F.R.G.	4,325	2,424	54.0	2,021	111	0.05	-2,313
Italy	2,151	1,470	68.3	697	12	0.01	-1,458
U.K.	1,785	1,556	89.0	144	5	0.03	-1,551
Finland	5,682	37	1.0	7,331	1,715	23.3	+1,678
Norway	497	129	25.9	1,608	446	27.7	+317
Sweden	5,721	77	1.0	8,530	2,886	33.8	+2,809
Japan	10,194	1,698	16.6	8,612	105	12.1	-1,593
Brazil	2,178	21	1.0	2,190	759	34.6	+738
U.S.S.R.	8,365	1,000	11.9	9,060	1,050	11.5	+50

Source: Drawn from _Pulp and Paper International_, Annual Review, 1982.

EXHIBIT 2-10

Imports, Exports, and Trade Balance of Paper and Board for Selected Countries, 1981

	Consumption (000 tons)	Imports (000 tons)	Imports as a Percentage of Consumption	Production (000 tons)	Exports (000 tons)	Exports as a Percentage of Production	Trade Balance (000 tons)
Canada	4,865	445	9.1	13,835	9,153	66.1	+8,708
U.S.	61,781	7,117	11.5	57,765	3,617	6.2	-3,500
France	6,231	2,174	34.8	5,148	1,091	21.1	-1,083
F.R.G.	9,619	3,818	39.6	7,828	1,961	25.1	-1,857
Italy	5,051	963	19.0	4,854	821	16.9	-142
U.K.	7,053	4,230	59.9	3,377	327	9.6	-3,903
Finland	1,342	64	4.7	6,135	5,247	85.5	+5,363
Norway	529	154	29.1	1,373	1,015	73.9	+861
Sweden	1,660	200	8.3	6,100	4,600	75.4	+4,400
Japan	16,767	490	2.9	16,980	673	4.0	+183
Brazil	3,223	234	7.2	3,297	298	9.0	+64
U.S.S.R.	8,905	1,000	11.2	8,955	1,050	11.7	+50

Source: Drawn from Pulp and Paper International, Annual Review, 1982.

EXHIBIT 2-11

International Pulp Trade Flow, 1966
(000 metric tons)

Importing Country \ Exporting Country	Austria	BLEU	Canada	F.R.G.	Finland	France	Italy	Japan	Netherlands	Norway	Spain	Sweden	Switzerland	U.K.	U.S.	U.S.S.R.	Yugoslavia	TOTAL
Austria			2.9	7.1	2.8		0.2			1.3		13.0	0.4		7.2	0.1	3.5	38.5
BLEU			17.4	3.8	54.0	3.6			1.3	21.4		120.6		0.1	33.3	1.0		256.5
Canada												4.2			41.6			45.8
F.R.G.	24.5	18.5	85.4		162.5	24.1	0.3		2.5	54.8		520.9	0.7		133.6	29.5	6.8	1064.1
Finland			0.7	0.1						0.1		3.8						4.7
France	1.1	16.0	34.2	18.6	180.8		0.5			72.7	0.3	437.8	4.1		99.4	35.2	2.0	902.7
Italy	71.8	0.1	79.1	12.3	186.6	30.4				43.7		336.3	2.3		126.8	12.6	21.6	923.6
Japan			236.4		26.9					0.1		6.5			329.0			598.9
Netherlands		12.4	40.0	19.6	201.2	7.7				72.7		174.2		0.9	38.0	0.6		567.6
Norway					8.8							65.1						73.9
Spain			0.4	0.1	50.7	23.0				36.4		102.1	1.8		9.9	0.3		224.7
Sweden					2.6					2.3					0.2			5.1
Switzerland	20.9		2.9	3.8	18.4	4.2	1.4		1.0	1.2		83.7		0.1	7.3			144.1
U.K.		11.4	300.8	4.6	661.7	16.3				509.0	0.3	1004.8			163.2	50.0		2747.5
U.S.			2701.4	0.4	96.7					0.7		172.0						2971.2
U.S.S.R.					106.5					2.0		27.4			23.4			159.3
Yugoslavia	7.4				0.4							7.1			6.2			21.1
TOTAL	125.7	58.4	3501.6	70.4	1760.6	109.3	2.4	--	4.8	818.4	0.6	3079.4	9.3	1.1	1091.1	129.3	33.9	

Source: The Pulp and Paper Industry, OECD, 1966-1967.

EXHIBIT 2-12

International Pulp Trade Flows, 1971
(000 metric tons)

Importing Country \ Exporting Country	Austria	BLEU	Canada	F.R.G.	Finland	France	Italy	Japan	Netherlands	Norway	Spain	Sweden	Switzerland	U.K.	U.S.	U.S.S.R.	Yugoslavia	TOTAL
Austria			8.9	12.1	7.9	0.1	0.4		0.7	1.0	0.1	28.9	0.1		17.1		0.1	77.4
BLEU			68.3	10.6	41.1	1.0	0.1		4.6	49.6		124.4			60.7	0.3	1.9	362.6
Canada								0.1				8.2			69.7			78.0
F.R.G.	13.0	27.6	324.3		133.5	34.9			6.5	73.8	0.1	582.3	2.6	0.7	193.2	4.7	13.1	1410.3
Finland			0.2	2.0								4.2			0.2			6.6
France	0.8	30.3	146.3	12.3	124.8		1.9		0.8	72.9	1.9	464.4	3.8	0.5	161.9	20.8	1.5	1044.7
Italy	68.9	6.6	188.3	5.3	131.6	27.9			0.4	25.0	1.3	289.5	5.2	0.8	150.7	8.5	8.9	918.9
Japan			347.3		9.5							1.1			285.2			643.1
Netherlands		17.4	109.5	8.6	140.9	4.9				80.4		147.7	0.2		69.9	1.2		580.7
Norway					9.9							88.4						98.3
Spain			26.7		55.5	26.1				23.2		97.7	0.1		31.5			260.8
Sweden			6.7	0.3	0.3					10.7					2.3			20.3
Switzerland	6.9	6.7		0.7	28.9	2.1	0.5		0.3	6.3		119.8			9.7			181.9
U.K.		0.5	331.1	1.1	392.5	4.3	0.7	0.2		418.0	7.8	694.1			217.3	3.0		2070.6
U.S.			3060.0	0.3	33.3	2.1				0.2		18.0		0.6				3114.5
U.S.S.R.			22.0							2.0		63.6			46.3			133.9
Yugoslavia	7.9		21.6		0.6					1.0		2.6			12.0			45.7
TOTAL	97.5	89.1	4661.2	53.3	1110.3	103.4	3.6	0.3	13.3	764.1	11.2	2734.9	12.0	2.6	1327.7	38.5	25.5	

Source: The Pulp and Paper Industry, OECD, 1971-1972.

EXHIBIT 2-13

International Pulp Trade Flow, 1976
(000 metric tons)

Importing Country \ Exporting Country	Austria	BLEU	Canada	F.R.G.	Finland	France	Italy	Japan	Netherlands	Norway	Spain	Sweden	Switzerland	U.K.	U.S.	U.S.S.R.	Yugoslavia	Africa	Asia	Latin America	TOTAL
Austria		2.7	27.6	19.4	6.9	3.0		4.3		5.9	3.5	42.9			24.4	1.9	33.3	25.2	3.2	4.9	209.1
BLEU			79.4	9.9	21.1	5.9		4.0	2.9	46.2	5.5	125.4			62.9			0.8		0.1	364.1
Canada						6.4		1.0				1.1			66.3			14.9		5.0	89.7
F.R.G.	11.4	27.3	430.1		174.3	35.6	0.8	4.0	3.6	104.2	16.6	694.5	1.0	0.1	264.3	28.4	14.6	5.8		22.4	1144.5
Finland	7.3		1.4							0.2		3.6		0.1	0.3	0.3					13.2
France	0.3	43.0	292.9	35.8	62.3		1.4	8.1	0.3	33.9	19.8	408.1	2.9	1.5	205.8	38.4	0.4	64.1		39.6	1258.6
Italy	83.0	3.5	287.7	26.2	90.8	49.4		0.1	0.8	35.5	10.2	305.4	5.9	1.1	199.8	19.0	18.5	17.2	0.6	11.4	1166.1
Japan			469.9		3.5							6.1		0.4	400.8			25.2	3.2	4.2	913.3
Netherlands		41.4	115.0	15.5	61.6	3.5	0.1			43.5		205.2	0.1	0.7	78.2	5.4				1.9	572.1
Norway					7.7							85.1			0.5					133.2	226.5
Spain			62.3	0.7	24.8	28.0				23.8		66.2		1.0	30.0						236.8
Sweden			0.1	0.1						16.5					0.5						17.2
Switzerland	3.1	5.8	5.4	24.3		6.8				5.8		115.6		0.1	13.9			0.1	0.1	0.3	181.3
U.K.		1.4	525.1	1.4	262.6	1.1	0.7	0.1	0.1	270.3	17.5	657.6			246.8	7.4		201.5	0.8	5.5	2199.9
U.S.			3245.3	0.6	11.9			0.3				46.0	0.2	0.8				25.0		0.2	3330.3
U.S.S.R.					121.9					7.3		52.9			20.4						202.5
Yugoslavia	23.6		15.2		8.7		0.2			0.3	0.2	16.0			13.1						77.3
Africa			32.8		9.6	8.1	0.2	6.0		0.8	0.5	35.9			115.2						209.1
Asia			223.2		56.5	7.1		95.8		6.3	0.5	81.5			198.1						669.0
Latin Amer.			67.5		7.7			0.8		21.2	0.6	21.7		0.4	253.0						371.1
TOTAL	128.7	125.1	5880.9	133.9	931.9	149.9	3.4	124.5	7.7	621.4	74.9	2970.8	10.1	6.2	2194.3	100.8	66.8	446.6	7.9	228.7	

Source: The Pulp and Paper Industry, OECD, 1976-1977.

EXHIBIT 2-14

International Pulp Trade Flow, 1981
(000 metric tons)

Importing Country \ Exporting Country	Austria	BLEU	Canada	F.R.G.	Finland	France	Italy	Japan	Netherlands	Norway	Spain	Sweden	Switzerland	U.K.	U.S.	U.S.S.R.	Yugoslavia	Africa	Asia	Latin America	TOTAL
Austria		0.1	23.4	13.1	10.5					4.6	1.8	32.0	0.1		31.5	32.9	12.2	2.5		8.0	172.7
BLEU			95.6	16.4	31.2	6.8	2.8		7.1	46.2	12.1	73.3			86.2	4.1		2.2		24.7	408.7
Canada								2.0							79.0					0.1	81.1
F.R.G.	24.1	26.1	532.2		324.0	38.4	3.6		1.7	93.5	49.8	589.4	6.2	1.0	430.8	61.7	4.4	12.1	0.8	152.8	2352.6
Finland	3.8	0.1	11.9							0.1		7.6			0.1			9.1		1.1	33.8
France	4.0	53.0	359.0	41.0	160.0		4.0			35.0	41.0	427.0	9.0		256.0	84.0	3.0	46.0		98.0	1620.0
Italy	122.1	6.2	378.0	16.0	162.1	58.9		0.1	0.5	32.7	30.1	252.8	5.4	0.3	236.6	29.8	20.4	23.5	0.8	29.1	1405.4
Japan			703.0		15.0						3.0	31.0			527.0			48.0	4.0	148.0	1479.0
Netherlands	0.2	39.5	113.2	16.7	73.6	2.0				23.5	7.1	125.1		2.4	120.0	16.8		4.9		21.2	566.2
Norway				0.2	31.8						7.7	78.3		1.0	0.1			5.4		151.6	277.9
Spain	0.9	0.2	60.9	0.2	20.6	25.2				7.2		55.9	0.1		42.7	3.2		0.3	0.3	10.3	228.0
Sweden			21.3	0.7	11.0					18.4				0.2	2.4			0.3	0.1	14.5	68.9
Switzerland	11.6	17.1	31.7	4.4	41.9	1.4				3.3		120.0			16.6	15.9		157.1		4.9	253.1
U.K.		4.2	274.7	1.6	275.8	6.0			0.6	135.9	14.9	379.4			260.5			96.7	1.0	66.2	1593.8
U.S.			3447.9		19.8			3.0		0.5		14.6								121.9	3704.4
U.S.S.R.			1.0		184.4		0.4			43.3		43.9		0.4	1.6						275.0
Yugoslavia	70.7		29.0	2.4	7.2	1.0	2.5			2.6	0.2	29.8			19.6						165.0
Africa	3.4	0.5	58.1	0.5	24.9	8.0	0.2				2.6	56.9		0.5	139.2						294.8
Asia			363.4	1.5	124.0	5.0	0.4	92.0		0.2		179.2	0.1		508.0						1273.8
Latin Amer.			98.0	0.5	4.2					22.7		6.0			364.5						495.9
TOTAL	240.8	147.0	6602.3	115.2	1522.0	152.7	13.9	97.1	9.9	469.7	271.6	2502.2	20.9	5.8	3122.4	248.4	40.0	408.3	7.0	852.4	

Source: The Pulp and Paper Industry, OECD, 1981-1982.

EXHIBIT 2-15

International Paper and Board Trade Flow, 1966
(000 metric tons)

Importing Country \ Exporting Country	Austria	BLEU	Canada	F.R.G.	Finland	France	Italy	Japan	Netherlands	Norway	Spain	Sweden	Switzerland	U.K.	U.S.	U.S.S.R.	Yugoslavia	TOTAL
Austria		1.0		11.0	10.5	0.6	2.2		3.7	1.7		16.7	4.0	0.7	1.0	0.1	0.3	53.5
BLEU	6.6		4.3	21.1	109.1	36.3	4.9		135.5	27.2		91.8	0.3	10.2	16.1	1.2	2.5	467.1
Canada	0.1			0.2	0.1			0.3	0.1			0.5	0.1	1.2	179.3			181.9
F.R.G.	144.8	42.9	73.3		553.0	58.7	36.3	3.6	199.4	153.0	0.4	505.3	3.3	2.3	162.9	0.2	1.9	1941.3
Finland				0.4		0.1			0.1			0.6		0.8	0.1			2.1
France	7.2	48.0	5.9	48.1	190.4		15.1	0.1	39.3	49.3	0.6	191.7	1.6	5.8	46.7	1.6		651.4
Italy	21.3	1.3	23.6	8.3	52.2	8.6		0.1	13.2	1.7		54.9	1.5	2.1	60.4		10.1	259.3
Japan		0.1	17.5	0.7	0.1							0.3	0.1	0.5	9.9			29.2
Netherlands	14.8	67.4	31.0	55.2	155.9	7.7	0.6			43.7		126.7	1.0	4.2	31.0		1.3	540.5
Norway	0.1			1.7	19.7				0.4			10.6		0.6	2.9			36.0
Spain	0.3	0.5	2.7	3.9	43.4	4.9	18.6		5.3	7.7		34.0	0.5	1.6	21.7		0.5	145.6
Sweden	0.5	3.3		12.9	9.0	0.1	2.6		2.8	0.8			0.1	1.0	3.0			36.1
Switzerland	18.9	0.5		11.3	16.8	3.8	2.6		3.7	9.2		34.2		3.4	2.3			106.7
U.K.	13.4	2.5	552.6	8.1	460.4	3.3	2.1		70.8	137.8		384.6	0.8		236.8	0.4		1873.6
U.S.	1.2		6173.7	2.7	287.2	2.3	1.3	45.7	0.5	6.3		34.9	0.1	2.8				6558.7
U.S.S.R.					195.1	1.2		0.3		3.0		11.9		0.1				211.6
Yugoslavia	0.7			0.5	0.4	0.2	10.4					0.6	0.1		3.6			16.5
TOTAL	229.9	167.5	6881.6	186.1	2103.3	127.8	96.7	50.1	474.8	441.4	1.0	1499.3	13.7	37.3	777.6	3.5	16.6	

EXHIBIT 2-16

International Paper and Board Trade Flow, 1971
(000 metric tons)

Importing Country \ Exporting Country	Austria	BLEU	Canada	F.R.G.	Finland	France	Italy	Japan	Netherlands	Norway	Spain	Sweden	Switzerland	U.K.	U.S.	U.S.S.R.	Yugoslavia	TOTAL
Austria		0.5	0.1	14.7	20.9	0.8	5.5		3.8	2.1		19.7	12.5	3.1	2.5	3.3	0.9	90.4
BLEU	11.6		9.0	84.2	113.7	103.0	2.3		154.8	34.2		119.1	0.4	13.6	42.0	3.7	12.8	704.4
Canada	0.1			0.3	0.4	0.2		1.8	0.7			0.5	0.1	1.8	250.1			256.0
F.R.G.	197.6	86.2	57.2		661.3	170.6	82.9	1.3	265.7	213.0	1.9	630.8	8.8	3.3	382.6	17.9	7.3	2788.4
Finland	0.6	0.2		1.2		0.1			0.2		0.3	1.2	0.1	1.3		8.0		13.2
France	2.3	96.2	26.8	190.2	249.3		25.9		55.9	77.7	6.0	286.6	1.2	15.1	117.8	6.1		1157.1
Italy	8.8	6.1	2.0	48.3	25.5	21.5		0.1	5.5	2.4	1.9	84.9	2.1	0.7	195.5		14.6	419.9
Japan	0.1		52.6	5.2	3.7	0.4						2.3		1.4	32.5			98.2
Netherlands	19.0	91.1	24.0	154.6	128.6	19.3	4.7	0.1		52.6	0.1	174.8	0.2	2.9	63.9	3.3		739.2
Norway	0.2			3.7	15.0				0.4			26.3	0.1	1.1	2.1			48.9
Spain	0.6	1.0	18.1	6.5	43.0	8.3	5.1		3.3	14.2		24.4	0.7	2.8	42.9			170.9
Sweden	0.7	1.3		16.5	39.7	0.2			4.9	2.9			0.3	2.5	4.0			73.0
Switzerland	59.5	2.4		17.8	62.8	3.3	2.4	0.1	1.8	22.1	0.5	72.6		7.0	2.7			255.0
U.K.	35.0	4.1	646.6	26.4	774.4	4.2	1.7	0.2	45.0	193.7	1.6	604.3	2.0		269.0	2.9	0.6	2611.7
U.S.	0.8		5994.5	2.8	301.5	3.7	0.1	1.9	0.4	3.5		25.1	0.3	3.0				6337.6
U.S.S.R.	5.8	0.1	0.8	2.8	223.1	24.7			2.0	52.0		110.1		1.2				422.6
Yugoslavia	18.8	0.2		14.3	4.2	1.0	38.2	0.1	0.4		0.2	5.8	0.2	1.0	14.6			99.0
TOTAL	361.5	289.4	6832.1	589.5	2667.1	361.3	168.8	5.6	544.8	670.4	12.5	2188.5	29.2	61.8	1422.2	45.2	36.2	

Source: The Pulp and Paper Industry, OECD, 1971-1972.

EXHIBIT 2-17

International Paper and Board Trade Flows, 1976
(000 metric tons)

Importing Country \ Exporting Country	Austria	BLEU	Canada	F.R.G.	Finland	France	Italy	Japan	Netherlands	Norway	Spain	Sweden	Switzerland	U.K.	U.S.	U.S.S.R.	Yugoslavia	Africa	Asia	Latin America	TOTAL
Austria		0.8	1.2	35.2	25.6	1.6	9.0		3.5	1.9	0.1	21.3	21.8	3.2	0.8	5.5	1.5				133.0
BLEU	17.3		27.8	175.7	68.4	130.8	22.6	0.7	162.8	16.2	1.3	132.1	4.3	20.4	35.7	8.5			0.2	26.8	851.6
Canada				0.2	0.1	0.2		0.3	0.3			0.8	0.1	1.5	526.6			0.1	0.2	4.4	534.8
F.R.G.	259.8	108.8	100.1		533.4	250.6	138.5	16.0	280.1	166.0	8.0	632.7	63.1	12.1	232.4	74.1	6.8	0.2	1.6		2884.3
Finland	0.9	0.1	2.7	2.5		0.3	0.1		0.4	0.3		10.7	0.7	2.8	0.3						21.8
France	14.8	120.2	42.0	379.7	266.1		64.7	5.0	78.1	83.8	35.1	347.5	10.7	47.8	98.7	1.1		0.2	0.3		1595.8
Italy	28.2	9.6	1.8	52.7	18.1	26.1		1.6	7.3	0.8	6.4	96.4	9.4	0.9	48.8	0.9	19.1		1.5		329.6
Japan		1.6	64.9	0.1	65.2	0.2								1.3	26.5						159.8
Netherlands	28.1	143.1	32.2	249.9	141.6	51.2	7.8	0.5		196.2	3.7	237.6	1.9	9.8	45.9	7.6	0.2	0.3	0.1		1157.7
Norway	2.4	0.2		3.3	36.5				1.4			54.2	0.1	1.4	4.7						104.2
Spain	3.2	1.2	26.9	22.8	61.1	11.0	3.2	0.1	1.4	11.6		42.0	0.3	3.8	59.3	6.7	0.1	2.0		0.7	257.4
Sweden	3.5	0.5		18.3	47.8	0.8	0.1		1.0	8.9			0.7	2.2	6.3						90.1
Switzerland	73.9	1.5		39.6	41.5	7.4	3.5	1.7	4.7	13.4		66.5		6.5	3.5	1.0	0.1				264.8
U.K.	47.4	19.2	600.7	71.8	1012.0	33.1	17.4	1.3	68.7	610.5	3.9	763.4	2.6		188.7	2.4	0.5	0.1	4.5	2.1	3450.3
U.S.		3.1	6196.3	0.5	71.0	4.9	0.1	0.1	2.9	7.8	0.1	42.9	0.6	2.6						4.5	6337.4
U.S.S.R.	0.3				338.1	29.8	0.5		5.1	35.5		50.0		1.7	0.1						461.1
Yugoslavia	18.0				8.0	1.0	6.8		1.1			5.6	1.5	1.1	0.8						43.9
Africa	75.4	4.8	19.7		108.6	49.4	16.2	6.4	20.3	58.5	33.7	124.3	1.0	50.7	205.8						774.8
Asia	78.6	1.7	214.2		192.1	13.6	11.2	540.6	8.3	56.0	17.0	153.6	1.7	19.4	279.2						1587.2
Latin Amer.	9.1	1.9	567.7		146.4	6.7	19.0	1.2	4.0	18.6	6.9	69.7	2.2	10.8	709.8						1574.0
TOTAL	660.9	418.3	7898.2	1052.3	3181.6	618.7	320.7	575.5	651.4	1286.0	116.2	2851.3	122.7	200.0	2473.9	107.8	28.3	2.9	8.4	38.5	

Source: The Pulp and Paper Industry, OECD, 1976-1977.

EXHIBIT 2-18

International Paper and Board Trade Flow, 1981
(000 metric tons)

Exporting Country / Importing Country	Austria	BLEU	Canada	F.R.G.	Finland	France	Italy	Japan	Netherlands	Norway	Spain	Sweden	Switzerland	U.K.	U.S.	U.S.S.R.	Yugoslavia	Africa	Asia	Latin Amer.	TOTAL
Austria		0.6		57.3	34.7	4.6	4.7		1.3	3.5		34.4	34.3	1.1	0.7	5.5	12.2				194.9
BLEU	26.2		36.2	227.5	84.5	158.0	34.0	2.3	150.4	34.4	2.3	178.2	2.7	15.9	16.9	9.3		0.2	0.1	2.4	981.2
Canada	0.3	0.2		1.2	3.3	0.5		0.4	0.7			2.0		1.1	366.3					0.6	376.0
F.R.G.	308.5	99.4	125.2		663.7	378.0	238.2	1.6	331.7	162.2	17.4	834.1	93.4	22.8	184.8	69.7	4.4	3.2	0.5	21.8	3560.6
Finland	2.0	0.4		6.2		0.3	0.1		2.8	0.5		26.7	0.3	1.6	0.7					0.1	41.7
France	24.0	175.0	40.0	534.0	366.0		104.0	10.0	93.0	66.0	67.0	499.0	34.0	65.0	39.0	7.0	3.0			1.0	2127.0
Italy	79.2	26.8	1.7	159.6	47.9	104.7		1.1	21.5	12.4	35.1	174.9	19.0	4.0	84.7	5.1	20.4	1.4	0.1	1.7	801.3
Japan			66.0	1.0	44.0					2.0		14.0		2.0	273.0				92.0		494.0
Netherlands	37.8	138.2	74.0	350.8	164.3	99.2	19.0	1.4		72.2	1.9	303.4	3.5	14.4	64.9	9.0		0.4	0.4	0.7	1355.5
Norway	3.9	0.2		6.7	64.6	0.2			1.8			82.9	0.1	2.9	3.5						166.8
Spain	8.8	1.9	14.6	34.6	80.7	33.4	19.1	0.1	3.5	19.6		51.0	1.3	6.4	47.6				0.3	10.5	333.4
Sweden	5.1	0.9		20.4	41.1	1.1	0.2		2.7	16.6			0.5	1.2	2.3			0.2			92.3
Switzerland	87.2	4.6	0.1	83.8	62.9	12.0	6.3	4.5	6.8	11.9	0.1	105.4		5.6	2.2	0.5		0.4		0.1	394.4
U.K.	55.1	40.8	704.8	204.4	1121.4	151.9	63.0	1.6	118.4	242.2	12.2	897.0	9.4		306.1	6.3		8.5	8.6	9.0	3960.7
U.S.		1.7	6979.8	14.6	109.3	2.4	0.8	0.5	3.3	15.9		34.2	0.3	4.2				0.2		30.8	7198.0
U.S.S.R.	2.2	6.3		28.3	584.6	14.0	0.8			48.6	2.2	36.6	0.4	0.6							724.6
Yugoslavia	70.7		29.0	2.4	7.2	1.0	2.5			2.6	0.2	29.8			19.6						165.0
Africa	85.3	5.1	13.0	33.4	214.3	68.0	35.7	7.0	19.6	47.9	58.3	219.3	1.1	54.5	272.3						1134.8
Asia	136.8	6.2	423.6	51.1	398.7	20.0	19.7	622.0	26.7	98.4	51.6	509.0	3.9	27.1	599.2						2994.0
Latin Amer.	6.4	8.2	731.0	8.4	252.4	12.0	14.9	2.0	6.1	14.3	20.9	136.5	1.4	6.9	848.9						2070.3
TOTAL	939.5	522.8	9239.0	1825.7	4345.6	1064.3	563.0	654.5	790.3	871.2	296.2	4168.4	205.6	237.3	3132.7	112.4	40.0	14.5	102.0	78.7	

Source: The Pulp and Paper Industry, OECD, 1981-1982.

EXHIBIT 2-19

Estimated Average Cost of Producing Bleached Kraft Paper-grade Market Pulp, First Half, 1982
(U.S. dollars/air dry ton)

	Softwood			Hardwood	
	Western Canada	Sweden	U.S. South	Brazil	U.S. South
Direct Costs					
Fiber	138	212	116	60	101
Chemical	37	--	45	36	32
Wages and salaries	55	85*	58	29	58
Energy	38	--	41	25	51
Other direct	37	85	47	--	46
Total direct	305	382	307	150	288
Capital Costs					
Interest	24	--	20	--	20
Depreciation	16	--	18	--	18
Other interest	12	--	17	--	17
Total capital and misc.	52	101+	55	177+	55
Total Cost	357	483	362	327	343

*The sum of the chemical, wages and salaries, and energy direct cost components.

+The sum of the interest, depreciation, and other interest capital cost components.

Source: Mead Corporation estimates reported in Pulp and Paper International, September 1982.

CHAPTER 3

COMPARATIVE ECONOMIC CONDITIONS AND STRUCTURES

As discussed in the preceding chapter on international trade, national differences in supply and demand conditions, production factor costs and availability, infrastructure, and industry structure are key determinants affecting international trade activity and patterns. The purpose of this chapter is to describe and analyze in much greater detail the comparative economic setting and structure of the US pulp and paper industry with comparative but less detailed sections on the Canadian and Japanese industries and with some references to Western European industries. After a brief overview of major producing and consuming nations, our discussion shifts to industry structure, industry cost components, and capacity.

BACKGROUND AND OVERVIEW

It is difficult to imagine any product with a more ubiquitous role in the civilization of developed, industrial economies than paper. So pervasive are the uses of paper in the business of economic growth that gross national product projections are widely used in the paper industry as a leading predictor of paper demand. As goes GNP growth, so go sales of paper. Considering the variety and range of paper products in daily life, it is not surprising that per capita consumption of paper may be used as an indicator of a country's economic development.

As Exhibit 3-1 shows, developing countries with low GNP per capita are clustered at the low end of the scale in per capita paper consumption. Most LDC's consume less than 20 kg of paper per capita per year. In contrast, the US annually consumes about 250 kg per capita. As income per capita rises, per capita paper consumption rises also, to an apparent ceiling of about 150 kg per capita, per year. Today it is unusual to see a country with per capita consumption much above 150 kg per capita per year, unless the country has fiber supply substantially in excess of domestic requirements.

The relationship between per capita consumption of paper and economic development as measured by GNP per capita is even clearer when GNP is adjusted to recognize the inadequacy of official exchange rate conversions. Exhibit 3-2 shows per capita consumption of paper and paperboard plotted against per capita GNP, adjusted using the ICP computation devised by the United Nations International Comparison Project (ICP) directed by Irving Kravis.

Countries with high per capita incomes must satisfy the correspondingly high domestic demand for paper by producing large quantities of pulpwood, pulp, and paper themselves, or by importing these materials from other countries more abundantly endowed with fiber resources and papermaking capacities. Exhibits 3-3 through 3-6 suggest an important pattern which will be addressed at several subsequent points in this study.

If we look at the top 20 consumer countries of paper and paperboard, we find 12 developed OECD countries, two CMEA countries, and six LDC's. Four of the six LDC's (Brazil, Mexico, Korea, and Taiwan) are newly industrializing countries (NIC's). The other two LDC's, the People's Republic of China and India, are the two largest populations in the world, countries whose total paper consumptions are large even at low levels of per capita GNP.

When we shift our attention to the 20 largest <u>producers</u> of paper, the rank ordering is similar to consumption, but different in some important particulars. If we next shift attention to the rank ordering of the top 20 pulp producers, we see yet another rank ordering, quite different from the first ordering of top 20 paper consumers.

These different rank orderings suggest several important patterns about paper consumption, production, and trade:

A large share of the consumption of paper in the top 20 consuming markets is maintained by international trade.

The US is in a class by itself in the scale of consumption and production of both finished product and raw material (fiber). The US produces and consumes more than three times the quantity of paper produced and consumed by its nearest rival, Japan. The US produces more than twice as much pulp as its nearest competitor in that category, Canada.

While most of the top 20 consuming nations must import paper and/or pulp to maintain their high level of consumption, only a few, such as Canada and Sweden, stand out as consuming significantly less paper than they produce. If we look at the pulp-to-paper phase of Exhibits 3-3 through 3-6, the same two countries (Canada and Sweden) stand out along with the USSR and Finland.

Several important countries (Japan, USSR, PRC) appear to have stable self-sufficient balances between paper consumption and domestic paper production.

While international trade in paper sustains consumption of paper in many large markets, it appears that trade plays a relatively larger role maintaining the flow of raw material (fiber) to the established paper industries of the top 20 consuming countries. Many of these top consumers do import paper to satisfy domestic demand, but apparently there is a definite preference to maintain a large domestic paper industry even where that industry must be fed increasingly expensive imported pulp. That is, the slopes of the lines connecting paper production and pulp production are steeper than the slopes of the lines connecting paper consumption and paper production. Import paper if you must, but the preference is for pulp imports to feed a domestic paper industry.

It is significant for this study that some countries, notably Japan and Western European countries, have built large paper producing capacities without a domestic fiber base sufficiently large or sufficiently cost-efficient to support these capacities. Later sections of this study will explore the implications of this imbalance more fully.

Few regions have an absolute excess of fiber supply over domestic needs, and in even fewer areas is the available fiber economically accessible. Although North and South America, Scandinavia, the USSR, Southeast Asia, and Africa all have significant forest resources, the lack of infrastructure and distance from the forest resource to population centers have in the past retarded the development of these resources for the pulp and paper industry outside North America and Scandinavia. However, as the long-fiber, softwood forests of Canada and Northern and Central Europe come under the pressures of acid rain, conservation/preservation, land withdrawals, overharvesting, urban and suburban sprawl, etc., there is always the persistent possibility that the long under-utilized forest resources of the USSR or tropical areas such as the Amazon Basin will play a larger role in world trade in fiber. However, recent experience in Brazil (Jari) and the USSR does not seem very promising. Though ultimate development and utilization of these more remote resources is certain, it probably will be a slow process.

The experience through the first half of the 1980's indicates very clearly that the infamous wood shortage so widely predicted in the recent past is not about to occur in the near future. Equally clear is the failure of LDC's to become the world's "wood basket." LDC's as a whole remain significant net importers of wood, and the rapid decrease in forest land in LDC's raises important, but as of now unanswerable, questions about the ability of LDC's to exploit the fifteenfold increase in yield per hectare (15 m^3 per hectare per year in "managed" tropical forest versus one m^3 per hectare per year in "exploitation" tropical forest) projected from investments in tropical forestry.

Large inventories of *Radiata* pine (often described as "too good to pulp and not quite good enough for dimension lumber") will be available over the next decade in New Zealand and Chile. Wood-rich countries, like most resource-rich countries today, are displaying an increasing reluctance to export forest products in raw form, preferring to add value by manufacture in the country where the resource is grown. This trend can be expected to challenge the established patterns shown in Exhibits 3-3 through 3-6. This reluctance to export raw materials could help shift trade further downstream from logs and chips to more highly processed pulp, paper, and converted products.

PRODUCTION AND CONSUMPTION

United States

The US enjoys the largest per capita consumption of paper in the world: about 25 percent greater per capita consumption than its closest rivals, Sweden and Canada. The US is also the largest producer and largest consumer of paper in the world. High levels of production and consumption in the US reflect the size and wealth of the US market, the relative abundance and low cost of both the fiber raw material and other crucial inputs (energy and water), and close proximity to another fiber-rich country, Canada.

The production of paper in the US is an old business and not a particularly innovative business. Papermaking is a very cyclical business and, today, very big business.

The American papermaking industry began before 1700 with a plant in Germantown, Pennsylvania. Modern, high production paper manufacture was made possible by the introduction of the Fourdrinier process over 100 years ago, and the basic technology of production remains the same today. Expenditures on research and development (R&D) are low in view of few process changes and the fact that product innovations tend to come from paper customers rather than paper manufacturers. In 1982, the US paper industry spent an average of 1 percent of sales on R&D, compared to a 2.4

percent all-industry average. In the chemicals industry, similar to the pulp and paper industry in some respects, spending was close to 3 percent of sales.[1]

The industry is cyclical in production. Per capita consumption declined between 1981 and 1982 in 12 of the 20 countries, as shown in Exhibit 3-7, with the largest declines among the heaviest users, and the industry is even more cyclical in profitability. However, paper is the ninth largest manufacturing industry in the US and the largest firms in the industry are familiar names from the Fortune 500: Twenty-nine paper fiber and wood products companies rank in the 1983 Fortune 500 by sales and four (Champion International, Georgia Pacific, International Paper, and Weyerhaeuser) rank in the top 100 US industrial firms by sales and asset value.

While the overall pattern of paper production and consumption in the US is clear--per capita consumption is high because of high GNP, the industry is large because of abundant factors of production and a large domestic market--some of the more interesting trends are buried in the aggregate data. The _kinds_ of paper consumed change over time and the balance of production is shifting between regions. Some of the most attractive success stories in the industry today are stories of acquisitions and of paper converters specializing in new products.

Summarizing the very aggregate trends in paper use over the last 60 years, it appears that from 1920 to 1950 a major shift took place, increasing the relative importance of paper consumed as paperboard and sanitary paper products. Since 1950 most categories of per capita end use have more or less maintained the share reached in 1950. Two exceptions are worth noting--newsprint has continued its long-term decline in share of end use paper consumption and the combined categories of groundwood and writing paper have experienced a significant increase in share--probably reflecting the increased use of paper in photocopying and the "information society."

The growth rate in US demand for paper products is forecast by Jaakko Poyry, the Finnish consulting firm, to be lower than growth rate in demand in Western Europe and much lower than the rate of demand in LDC's. However, in contrast to Europe, where paper industry rationalization and pruning of pulp and paper capacity have been taking place, US papermakers seem to be showing an interest in increasing capacity in order to serve world markets via exports--especially Western European markets.

Although in the early 1980's financial returns in the industry were too low to encourage much construction of new or "greenfield" pulp and paper making capacity, investment in new plants and/or expanded capacity of existing plants over the last decade has been substantial. Both new and incremental additions to US capacity have been greater than additions to capacity in most other areas of the world. However, additions to capacity in the US are weighted heavily in favor of the US South. A 1981 survey by the FAO showed that for the period 1981 to 1991, 58 percent of planned US paper capacity additions and 78 percent of planned US pulp capacity additions were in the South.[2]

More recent surveys of capital spending plans by Pulp and Paper confirm this bias toward the South: 65 percent of capital spending in the period 1983-86 will be in the South and three of the top five states in pulp and paper capital expenditures are in the South. Other surveys corroborate the Pulp & Paper and FAO surveys. The API predicts that the West will lag far behind the South in absolute capacity expansion through 1986, and in terms of new mills (as opposed to incremental capacity in existing mills) two of the three planned US plants will be in the South.

Canada

The Canadian industry faces a turning point in its long history of fiber and paper production. The next few years may be among the most crucial in the determination of the future of Canada as a world class competitor.

Canadian per capita paper consumption--in excess of 170 kg per person per year--is relatively high, as might be expected of a wealthy country with abundant fiber supplies. Several factors point to the heavy reliance of the Canadian paper industry, and indeed much of the Canadian economy, on exports. Canada's pulp and paper industry is very large in absolute terms, being the third largest paper producer and the second largest pulp producer in the world. The industry's importance to the total Canadian economy is also great, in 1978 generating 7.9 percent of manufactured goods shipped, 8.6 percent of total wages and salaries paid, and 8.8 percent of total industrial value added. This makes paper and allied industries the third most important Canadian manufacturing sector behind food and beverages, and transportation equipment.

Given the relatively small size and wide geographic dispersion of the Canadian domestic market, exports of pulp and paper are a virtual necessity. Canada's largest and most stable export customer is the US, with nearby large urban markets for paper, especially newsprint. Other major trade partners for Canada are Europe, which imports pulp and newsprint from eastern Canada, and Japan, which imports chips and pulp from British Columbia.

As a supplier to the US, Canada is gighting against the increased paper capacity in the South, especially in newsprint and linerboard. These two products are commodities in which low fiber costs are primary competitive factors, and with Canadian fiber costs estimated at twice the South's, it is increasingly difficult for the Canadians to maintain their traditional market share in the huge newsprint markets of the east coast cities. The disadvantages Canada faces include a slower forest crop rotation, more expensive transportation from forest to the mill, sparser forest stands, and a less-developed infrastructure and technology for handling pulpwood, all of which contribute to the higher cost of providing fiber to pulp mills. Spruce bud worm infestations and, more recently, acid rain have been added to the list of uncertainties and disadvantages faced by the Canadian industry in the raw material stage.

Other disadvantages to the Canadian industry include older, more inefficient mills; high energy costs; and more expensive, militant labor both in the woods and in the mills. Total production costs to Canadian producers over time may be less than those of Southern producers because the older Canadian mills are largely depreciated and carry lower overhead costs. However, it is variable costs that determine the all-important operating rate when the industry is facing depressed markets, as has been the case during the past two years.

The increase in Southern capacity and aggressive moves by US companies into traditional Canadian markets have had a devastating effect on many Canadian producers. In the past three years the US has increased its newsprint capacity from 35 percent of its domestic needs to 45 percent domestic sufficiency, cutting into the Canadians' largest export item. Canada's exports to th US have been heavily weighted toward newsprint due in part to a longstanding gap in the US tariffs, which permit free entry of newsprint but impose duties on virtually all other types of paper products.

Canadian tariffs on paper imports are in most products much larger and therefore more protective than US duties, suggesting that Canadian paper producers may not be competitive even in their own domestic markets in many items.

In Canada's other fiber export markets, we also find increasing US penetration. Europe in the late 1970's and early 1980's began to turn to the US South as a medium- and long-range supplier of fiber for the Western European paper industry, as Scandinavian forests neared or exceeded maximum sustainable harvest rates. Even some Scandinavian producers, traditional fiber suppliers to Europe, are importing Southern pulp and pulpwood. Finally, in the Pacific Rim, we have seen Canada's share of chip exports, most of which are to Japan, declining as the US's share grows.

Since newsprint has been the traditional mainstay of the Canadian industry, accounting for 64 percent of 1980 Canadian paper production, one may wonder why Canadian producers have not diversified more into other types of paper and paper products. The Great Lakes region of the US paper industry, with its specialty paper production and older, established paper mills, has a counterpart region in Canada, north of the Great Lakes region. Some specialty products are made there by the same types of older mills that are found in the US. These specialty producers compete directly with most of the rest of the 36 percent of Canadian production which is not newsprint. As with most of the industry in the eastern part of Canada, the mills there are suffering from freight disadvantages from forest to mill and from mill to market. Canadian producers also face increasing fiber and energy costs; competition from newer, more efficient US mills; and a general lack of investment to keep the industry current. Given small Canadian domestic markets for specialty products and eager, highly integrated American producers willing to use their low-cost advantages of production, Canadian producers cannot always be competitive in their own domestic markets, even given tariff protection and an exchange rate advantage of a weak Canadian dollar and strong US dollar.

The Canadian forestry community is likely to explain Canadian problems in terms of division of responsibility for Canadian forests under an unsatisfactory provincial/federal split. (Private forests are insignificant in Canada except in some areas of Quebec and Eastern Canada.) This split responsibility has allegedly encouraged provincial governments to exploit forests to pay for provincially provided social services with accumulating backlogs of harvested but unreplanted forest.

Another major problem in the Canadian industry seems to be a lack of investment capital coupled with a high cost of building new, more efficient capacity. Returns in recent years have been so poor that Canadian paper companies were unable to afford to upgrade or rebuild capacity in order to become competitive with American producers. It is estimated that the costs of building a "greenfield" paper mill in Canada will exceed by $1500 per ton of capacity the costs of building the same mill in the US.[3]

Unless this trend toward "patching up" current capacity turns around, Canadian producers will soon reach the limits of improvements by such means as speeding up old machines and making them somewhat more efficient in other ways. If the Canadian industry cannot overcome higher operating costs and fiber costs to again become more competitive in international markets, the prognosis is not good for Canada's future.

Japan

Japan's paper industry is facing problems no less severe than Canada's but the structure and orientation of the two industries are drastically different. With the second largest paper capacity in the world, much of it in relatively new mills with modern technology and high productivity, the Japanese industry has traditionally been the sole supplier to its own domestic market.

Ranked ninth in world per capita paper consumption in 1982, Japan's aggregate consumption of paper and paperboard has risen from 4.5 million metric tons in 1960 to well over 17 million metric tons in 1982. Imports have risen from almost nothing in 1960 to over half a million tons in 1982, still only 3.5 percent of Japanese production. Imports in 1982 were more than offset by 740,000 tons of exports of higher unit value exports, mostly to East Asian countries. Comparing Exhibit 3-8 to Exhibit 3-17 we see that the gross pattern of end use consumption of paper in Japan is very similar to the pattern in the US.

The Japanese paper industry is characterized by a high level of import dependence on raw materials, a high level of recycling wastepaper (45 percent of the furnish to Japanese paper and paperboard production is waste paper and other fiber supply), and decreased international competitiveness caused by increases in energy and raw material costs. Although costs are increasing, price levels in the early 1980's have been depressed by slack domestic demand and strong domestic competition, catching Japanese producers in a cost-price squeeze. This squeeze has resulted in some unusual government action discussed in greater detail in the following chapter.

Although not yet an important importer of paper (this could change in the next decade) and never a major exporter, the pattern of Japanese imports and exports is shown in Exhibit 3-18.

Exhibit 3-19 shows the flow of raw material and products in the Japanese pulp and paper industry. As can be seen, about 40 percent of pulpwood and about 16 percent of pulp is imported, most of it from the US and Canada. Actual reliance on imported fiber may be somewhat less, however, since the Japanese industry uses more than 40 percent recycled fiber (wastepaper) in its furnish. Much of this is recycled domestically, and some is imported from North America. Exhibit 3-20 shows supply and consumption over time of pulpwood to the Japanese industry, illustrating Japan's increased reliance on imports.

The Japanese industry is also structurally depressed in that there is definite overcapacity in several product types, notably linerboard. Recent increases in the costs of chips and oil have brought the Japanese industry face to face with lower-cost North American producers. The same North American firms which supply fiber to Japan also are competitors in paper production. US and Canadian producers have been increasingly successful in penetrating the Japanese paper market, most effectively in commodities such as linerboard and newsprint, for which cheap virgin fiber and large economies of scale are important advantages. Exhibit 3-21 displays price indexes of pulpwood and chips over time, compared with the wholesale price index for Japan.

Two developments caused severe price depressions and reduced operating rates in the Japanese industry. The first was increased imports of paper and paperboard coupled with excess existing capacity in some areas. The second was new production capacity planned by domestic Japanese companies in an effort to improve their competitive advantage over other domestic and international producers ("pre-emptive capacity"). Depressed demand for business and industrial paper grades during the depression of the late 1970's and early 1980's caused Japanese pulp and paper industry leaders to petition the government for permission to form recession cartels. Restricted production stabilized falling prices in some cases, and an industrywide rationalization program is under way. This situation will be covered in more depth in Chapter 4.

The future direction of the Japanese pulp and paper industry is uncertain. Improvements in profits and operating rates in 1983 seem to be the result of the restrictions allowed by government. Efforts are being made to increase exports but exports are quite limited and concentrated.

Printing and writing papers, whiteboard, and newsprint account for 74 percent of exports. East Asia takes 87 percent of Japanese paper exports and two markets, Hong Kong and China, account for 50 percent of Japan's paper exports.

MITI's outlook for the long term of the Japanese paper industry recognizes its decreased competitiveness and predicts a growing share of Japan's paper consumption will supplied by imports. The US and Canadian industries, but especially the US West Coast industry, would, of course, like to hasten the realization of the MITI vision for the Japanese paper industry.

Summary

In a summary of trends in production and consumption of paper, certain trends apparently are well established and clearly visible. Pulp production tends to be located near accessible and appropriate forest resources and suitable production sites. But large supplies of fiber alone are not enough to guarantee a correspondingly top position in pulp production as shown in Exhibit 3-23. Paper production is usually located in small cities near the pulp resource (integrated with pulp mills) and, if possible, not too far from urban or business centers. Converter operations tend to be located closer to the market. Pollution problems and water requirements tend to keep pulp and paper mills rural. Transportation cost is significant (freight costs account for about 8 percent of the industry's total costs)[4] throughout the industry but freight sensitivity is especially important at the fiber end of the business because of the low value high weight, high volume of the raw material.

As currently overutilized sources of fiber become scarcer and more expensive, new sources of fiber become relatively more attractive. In addition to the well documented relative shift to the US South, wastepaper and the conversion of "exploitation" forest resources to "managed" forest resources in Oceania and some regions of the Third World could become more attractive as fiber sources. As some previously underdeveloped countries continue to experience rapid economic growth and increases in GNP, the centers of consumption of pulp and paper should also shift toward these new centers. The evidence of the last decade or two suggests that the "center of gravity" of economic growth is shifting to the Pacific Rim. Hence the emphasis in the next section will be on the US, Canada, and Japan, as the biggest current markets and suppliers on the Pacific Rim.

INDUSTRY STRUCTURE

The United States

The pulp and paper industry, contained within SIC Code 26, "paper and allied products," consists of four major four-digit SIC code titles:

2611 Pulp mills
2621 Paper mills, except building paper
2631 Paperboard mills
2641 Building paper and board mills

Pulp mills are the establishments which break down wood fiber from solid pieces of wood to a pulp suitable for the manufacture of paper products. In the US in 1977, there were 45 pulp mills, 42 of which had 20 employees or more. At that time, there were 16,200 employees working in these mills, and the major states in terms of employment in this industry were Washington, Florida, Georgia, and Maine, which together employed 60 percent of the total. The total value of shipments from these mills in 1977 was $2,091 million, $1,332 million of which was accounted for by bleached kraft pulp. Value added by manufacture for all pulp was $906 million.

Paper mills, excluding building paper, are establishments which use the pulp generated in pulp mills, usually in a liquid slurry form, to manufacture paper. Paper mills are sometimes combined with plants which also manufacture converted paper products, such as boxes, egg cartons, or any of literally thousands of other finished products. In the US in 1977, there were 339 paper mills, 287 of which employed more than 20 persons. One hundred twenty of these, or 42 percent, were integrated with a pulp mill on the same site, but these integrated mills were responsible for 72 percent of production, by value of shipments.

The remaining unintegrated mills were either dependent on market pulp to support their production or were contractually "tied" to the production of pulp mills in the US or Canada.

At the time of the last census report available, 1977, 127,000 employees worked in this segment of the industry, and 35 percent of them worked in Wisconsin, Maine, Pennsylvania, and New York. This industry segment showed impressive growth, partly due to inflation between 1972 and 1977, with an overall increase in the value of shipments of 98 percent, to a 1977 figure of over $12,000 million. One of the highest growth product segments in this category was uncoated freesheet papers, such as are used for dry copiers and typing papers, which increased 101 percent since 1972. Value added by manufacture for the paper mills segment was $5.4 billion, which was an 86 percent increase over 1972.

Industry 2631, paperboard mills, includes all mills which manufacture paperboard from pulping fibers. Paperboard manufacture is a variation on paper manufacture which layers successive webs of fiber together to form thicker, sometimes structural sheets. Examples of end products include food boxes and cartons, corrugated cardboard, and various other heavier paper products. This classification also includes converters which are integrated with primary paperboard manufacturers, as well as pulp mills which are integrated with paperboard mills, but are not separately reported. In 1977, there were 249 paperboard mills in the US, 232 of which employed 20 or more persons. Eighty-two of the 232, or 35 percent, were integrated with a pulp mill, although these 82 produced nearly 80 percent of the value of shipments, which totalled $7.124 billion for the industry. This was an increase of 71 percent over 1972's value of shipments, while value added by manufacture increased 54 percent in the same period, to $3.1 billion. The segments of this industry which showed the most significant growth were bleached converting paperboard, at 96 percent growth, and semi-chemical paperboard, which increased by 80 percent over 1972. In 1977, the industry had 67,600 employees, 35 percent of whom worked in Georgia, Louisiana, Alabama, and Virginia.

Structural Silhouette

> It is difficult to trace a firm silhouette of the paper industry because many companies have large investments in forests and many are also substantially engaged in converting; some companies sell pulp and some only make paper. In short, the harvesting of raw materials and the conversion into end products are often divisions of a given paper company.
>
> Strange, 1978

Several other differences can be added to the Strange quotation. Many paper producing firms with large investments in forest lands integrated forward into paper to maximize use of residues from wood products manufacture, while other paper firms are attempting to divest their forest holdings. Some paper producers are today primarily producers of solid wood products, while others have shifted their product mix toward more lucrative and less cyclical paper products. Some paper producers expanded horizontally into paper from related industries, while others have expanded into paper as a diversification strategy per se, and still others have integrated backward into paper from printing or publishing. Recent industry mergers and acquisitions have made it more difficult than ever to draw a single silhouette of a typical paper company.

Although industrial climates differ from country to country, and the pulp and paper industry assumes a different shape in each, certain factors are universal to the business and affect all companies engaged in the production of pulp and paper. Relative advantage depends partly on proximity to natural resources and markets, and partly on how well the individual company utilizes natural structural advantages and deals with structural difficulties relative to domestic and international competitors.

Nevertheless, there are some generalizations one can make about the structure of the industry:

* The industry is capital intensive and concentrated but
    * prices are below the price index for all industrial commodities.

  * the industry is not as concentrated as one might expect given the enormous level of capital outlay and environmental problems associated with a new facility.

* Large capital outlays have been associated with

  * outstanding increases in productivity,

  * large value added to a relatively few inputs, but

  * limited scope for further increases in economies of scale, and

  * relatively low expenditures for research and development.

* The industry is widely viewed as a "growth" industry

  * because paper consumption increases with GNP, but

  * is characterized by alternating periods of excess capacity and capacity shortage, hence

  * profitability is both lower and more cyclical than the average for all manufacturing.

* Part of the industry produces a commodity that sells at international commodity prices, but some of the industry produces heavily advertised and highly differentiated consumer products.

* Upstream segments of the industry (i.e., pulp and paper mills) are located near the raw material but converting facilities are located near high income markets. (I.e., in 1976 the South, with 29 percent of US personal income, produced 67 percent of US pulp, 50 percent of US paper and board, but only 26 percent of US converted paper products, while the North with 52 percent of US personal income produced 16 percent of US pulp, 37 percent of paper and board, and 62 percent of converted paper products.)[5]

* Companies in the industry increasingly derive larger shares of sales from nonpaper products, such as nonwoven textiles, flexible packaging, etc.

Concentration

> The critical mass for paper companies has increased markedly since World War II. One reflection of this phenomenon is the steady reduction in the number of independent corporations. Between 1950 and 1970, for example, the total of primary companies shrank from 570 to 350, or a decline of 40 percent. Over this same span of years, production increased by about 155 percent.
>
> Strange, 1978

A "small" paper company in 1972 had sales between $5 million and $40 million. But in 1972 of the more than 200,000 manufacturers in the US, 99 percent had sales less than $20 million. In 1983, the largest paper producer in the world, International Paper, had total sales of $4.4 billion (about $3.3 billion from paper) and ranked 80th in the Fortune 500. So a small paper company is not really very small, and although a large paper company can rank in the top 100 of the Fortune 500 (with a boost from nonpaper production), it is still way down from the $25 billion to $90 billion level of sales achieved by really big petroleum, auto, and chemical companies at the pinnacle of the Fortune 500.

In 1977, the two-digit SIC industry "paper and allied products" comprised 17 subindustries and shipped $52 billion worth of product. Although much more concentrated than its sister industry "lumber and wood products," the paper industry was less concentrated than "food and kindred products" and "textile mill products" and much less concentrated than "primary metals" and "transportation equipment." To illustrate, in 1977 in the "textile mill products industry" there were 13 four-digit subindustries where the value of shipments of the four largest companies in each industry accounted for more than 40 percent of the total industry value of shipments. The total value of shipments in these 13 subindustries accounted for 35 percent of all shipments in the textile mill industry. In the paper industry there were only seven subindustries where the value of shipments of the four largest companies in each industry accounted for more than 40 percent of the total industry value of shipments, and the total shipments in these seven subindustries was only 22 percent of the total value of paper industry shipments.

Guthrie (1972) discussed the difficulty of identifying and measuring concentration and concluded that there had been no significant trend in concentration in the paper industry in the time period 1947 to 1967. This stability was in contrast to the marked increase in concentration among the largest US manufacturing firms in the same period. He noted that concentration was highest in pulp and sanitary paper products industries for reasons of economies of scale and product differentiation by advertising. Similarly, Strange (1978) maintained that the level of concentration was virtually unchanged in the 1960 to 1974 period if one looks at value of shipments.

The 1977 Census of Manufactures shows a decrease during 1963 to 1977 in the number of companies in six of the nine four-digit industries accounting for more than 85 percent of the value of shipments in the "paper and allied products" industry, SIC 26. The apparent decrease in the number of pulp mills from 56 to 26 between 1967 and 1977 was largely explained by a change in definition and classification in 1977. The total number of companies in SIC 2621 (papermills), 2631 (paperboard mills), 2643 (bags), 2647 (sanitary paper products), and 2651 (folding paperboard boxes) decreased from 1458 to 1277 but the percent of value of shipments accounted for by the largest four firms increased only slightly in the other two. In three of the nine industries--2641 (papercoating and glazing), 2649 (converted paper products, n.e.c.), and 2653 (corrugated and solid fiber boxes)--the number of companies increased from 1144 to 2100 between 1963 and 1977, and the four-firm level of concentration fell or remained stable. These three industries accounted for almost 25 percent of SIC 26 shipments.

Of course, mergers and acquisitions in the late 1970's and 1980's could change these concentration ratios. Kline's Guide to the Paper Industry 1980 lists more than 60 acquisitions in the 1975-1979 period, and in the 1980's mergers and acquisitions continued to excite the attention of industry analysts--as did the disposal of substantial paper making assets by some companies, e.g., Continental Can, Time-Life, Crown Zellerbach, and International Paper.

Capital Intensity, Scale, and Integration

Three outstanding features of the economic structure of the US paper industry are its capital intensity, scale, and integration. Simply stated, the industry is very capital intensive; it is characterized by large economies of scale and high degrees of backward and forward integration. While there are some interesting and recent examples of successful companies basing growth and profitability on concentration in specialized products and achievement of extraordinary increases in operational productivity (James River), economies of scale and capital intensity dominate the character of the upstream paper industry.

One good measure of capital intensity is dollar assets per employee. Exhibit 3-26 shows that the pulp and paper industry ranks well above the manufacturing industry average in assets per employee. The pulp industry employs seven times more assets per employee than the average of all manufacturing. The assets per employee in pulp and paper mills have also grown more than the average for all manufacturing since 1963.

While papermaking is by nature a capital intensive industry, pollution abatement investment to satisfy state and federal water standards, air quality and solid waste requirements has further increased the capital requirements of the industry. (Recall that in Exhibit 3-12 more than 10 percent of capital spending planned for 1983 to 1985 in South Carolina, Alabama, Washington, Mississippi, and Michigan was "environmental investment.") Today the industry can be called very capital intensive with a one-to-three ratio of sales to plant investment.[6]

This capital intensity has become a key issue in the industry: As interest rates remain high, inflation and the cost of capital emerge as prime consideration in the decision to build a new mill. About 60 percent of North American paper machine capacity was installed before 1960, 25 percent from 1960 to 1970, and only 15 percent since 1971.[7]

Although forecasts of increased paper demand call for an increase in chemical pulping capacity of some 40 million tons in the 1980's (45 percent of total world chemical pulp production in 1982), there is a real question of whether new investments will be profitable at current levels of capital costs and pulp prices.

> The investment cost of pulp mills that started up in the early 1970's was about US$ 350/ton of annual capacity. By the late 1970's the unit cost of investment had risen to US$ 850. In the case of a big Brazilian mill started up in 1978, the investment of US$ 1500/ton of annual capacity, including the infrastructure, was astronomical. But current investment cost estimates are

> US$ 1800 to US$ 2000/annual ton of capacity in the industrial countries, and up to US$ 3000 to US$ 3500/ton in the developing countries. Rising costs mean that the price of long-fiber pulp should now be about US$ 750 to US$ 775/ton c.i.f. Rotterdam, to make new capacity profitable. In 1985, the price required will be about US$ 850 to US$ 900/ton.[8]

Pulp and paper mills are large and have been getting larger over time. In 1980 the average size of Ontario and Quebec's approximately 100 "old" newsprint machines was 65,000 tons per year, while the average size of newer machines in the South was 125,000 tons per year. In 1981 Publishers Paper, MacMillan Bloedel, and North Pacific each brought on stream 190,000 ton newsprint projects. Since 1981 North Pacific's two machines have achieved an annual capacity well above 400,000 tons per year. Kraft linerboard machines are even larger. In 1980 Union Camp brought a new 350,000-ton per year machine on stream in Alabama and in 1981 International Paper brought on stream a machine with 460,000 tons per year capacity.

During the 1970's one theme in restructuring the global paper industry was a movement to larger scale machines. Total world production increased 3 percent per year or 45 million metric tons in the decade, but the increase came about in the following manner: 2,195 machines with an average capacity of 7,750 tons per machine per year were shut down and 924 new machines were installed with an average capacity of 45,500 tons per year. Most of this restructuring was accomplished by rebuilding existing machines to improve capacities at an average of 14,000 tons increase in capacity per machine per year.[9]

One of the principal differences in the structure of the pulp and paper industries between countries is the degree of integration. In the US the typical company is integrated backward to timber ownership and forward to conversion. In Canada forward integration tends to be truncated due to the export orientation of the industry and small domestic market. Backward integration in Canada is provided by long-term timber harvest licensing arrangements between companies and provincial governments with stumpage prices tied to product prices. In Japan companies are integrated forward, but few own and supply a significant portion of their own timber resources.

Exhibit 3-27 shows fiber self-sufficiency and timber ownership of major US forest products and paper companies. Fiber self-sufficiency is widely held to be an important key to success in the forest products industry. More recently integration in other directions has been stressed.

Timber ownership is relatively less important in explaining growth, success, and profit in paper than in solid wood products simply because the cost of wood accounts for only about 20 to 30 percent of total manufacturing costs in paper compared to 60 to 70 percent in lumber and plywood. Nonetheless, timber ownership is an important dimension of the business. In addition, earnings from firm-owned timber harvest are taxed at the lower 28 percent capital gains rate. Furthermore, timber prices, and therefore chip and pulp market prices, are expected to rise faster than inflation. Access to owned supplies, access to wood on the long-term fixed-price-purchase-agreements prevalent in the South, or access to indexed Canadian stumpage prices also helps firms avoid the uncertainty of fluctuating and rapid price increases most vividly characterized by government timber sale contracts in competitive bidding west of the Cascades.

> At least 30 publicly owned companies have acquired substantial amounts (at least 50,000 acres) of timber-land since 1973, including 15 of the largest forest products and paper companies, most of which have publicly announced programs for the continuing purchase of timberlands. Fifteen other firms, including nine diversified companies, also have acquired considerable timber holdings....Since 1976 there have been 28 acquisitions completed in the US [18 of which involved timber assets].[10]

Investment to secure supply is by no means limited to the US industry. Japanese companies in the 1970's invested heavily overseas in plantation and afforestation projects to secure supply. But recently stagnant demand and the depressed financial conditions of the paper industry in Japan have made it both more difficult and less imperative for Japanese companies to continue developing overseas sources of supply.

Additional factors in other markets have suggested other kinds of integration. Increased energy and wood costs in Scandinavia have encouraged Scandinavian producers to try to move out of low-value, high-energy paper production into lower energy, higher value paper product production. This trend implies a move to integration of Nordic and West European industries. The horizontal integration successes of James River and promise of production efficiencies in integrated forest product and paper operations suggest other motives and types of integration. But the 1981-1982 recession brought about pruning and divestitures (e.g., Crown Zellerbach, International Paper, and Scott) that might be offered as counter-examples of integration. This theme of integration will be more thoroughly discussed in the strategy chapter.

The natural presumption of the industry described so far--very capital intensive, large economies of scale, and highly integrated--is that operating rates would be very critical to profitability. This is certainly true in the case of the paper industry. As observed already the industry is cyclical not so much in production as in profitability. In fact the paper industry is clearly less cyclical than lumber products and appears in Exhibit 3-28 to be less cyclical than durables and all textile mill products. This suggests what paper manufacturers know: Profitability is inextricably tied to operating rates. Any operating rate below about 90 percent constitutes a "buyer's" market, in which price cutting and discounts are offered to customers simply to keep the machinery in operation. At rates above 92-93 percent, spot shortages and tight supplies begin to occur, and prices rise accordingly. With such a narrow range of operating possibilities it is clear that profitability, so closely tied to operating rates, tends to be elusive in this industry.

Profitability is indeed quite variable and is closely linked with the timing of business and economic cycles in each country. What may not be immediately apparent is the delicate balance between overcapacity and shortage the industry must maintain, both in the domestic industry and in international markets. Because each additional paper plant or pulp mill is such a large incremental addition to capacity, and since there is no advantage of running the mill at anything but full operative capacity, each new mill must immediately not only produce but instantly market its full output.

Given that a new mill requires at least three years from financing to the start of production, the risks involved in such an investment are enormous. Not only may the markets change in the intervening time, but domestic or foreign competitors may build competing mills which would create sudden overcapacity. Industry observers suggest such an overcapacity now exists in the eastern US-Canada markets where southern US producers have built large incremental newsprint capacities to take advantage of domestic US markets which Canadians have dominated for years. Although the southern producers are now low-cost producers for those markets in every way, if all other factors were equal, the older Canadian mills might still have lower total costs (although not variable costs) than the newer southern competition because their older mills are fully depreciated, while the new mills must pay for enormously high fixed costs.

It is these kinds of tradeoffs between over- and under-capacity which are currently causing established markets to be less stable than expected. Last year's low-cost producer may not remain so this year, and predicting trends for the future, so important to this industry, become nearly impossible. As mentioned in the last section, the pulp and paper industries in Japan and Canada are facing some very difficult structural problems within their domestic frameworks, and international markets and patterns of production and consumption are changing as well.

Japan

Concentration

The Japanese pulp and paper industry is quite concentrated. Although there are many small-scale Japanese papermakers, making traditional and highly prized washi by hand, the top industrial Japanese paper mills are world-class operations in scale, efficiency, and product quality. The bulk of the machines owned by top companies are between five and 15 years old, which is about the same as the industry average in the US. Attempts have been made to increase capacity of existing machines and to modernize older facilities, similar to trends in the US and Canada.

Recent industry rationalization, assisted by the Japanese government, can be expected to increase efficiency further, in part by scrapping the most obsolete and least-efficient capacity of each firm. This industry rationalization can also be expected to further increase concentration in many segments by encouraging mergers and acquisitions.

Exhibit 3-30 shows a slow but steady trend in concentration in all pulp and paper segments, but the trend is reversed in the paperboard segments of linerboard, corrugating medium, and white board. These product lines are the very areas in which overcapacity is the biggest problem in Japan, and this structural imbalance has been a fact of life in the Japanese industry for some time.

Barriers to entry in Japan are the same as barriers to entry in this industry everywhere: high capital requirements, cyclical and often inadequate returns, domestic and world overcapacity, and high cost of capital. Added to these barriers in Japan are the difficulty in obtaining scarce raw materials, the relatively higher cost of energy (Japan imports virtually all her petroleum), and the huge amount of level land required for paper manufacture--a high cost feature of this industry in a country where industrial land is very scarce. Year-round clean water is another concern, as is a very high level of pollution-control requirements, comparable to those in the US.

Added to these obvious barriers to entry is the fervent and traditional competition within the paper industry in Japan. The five largest firms, most originally family owned, have been in competition with each other for many years, and intense rivalry has been the rule. Although they sometimes work together in ways that seem collusive to Americans, price cutting and pre-emptive capacity installation are common competitive tactics, which may help account for the low level of profits often shown by Japanese industry. The strategy of installing capacity before demand truly materializes to prevent competitors from planning new capacity is common in Japan and may be one reason for the overcapacity the industry now faces, as well as part of the reason for industry concentration in some segments.

Japanese paper companies are large by Japanese and international standards. Sales of Oji Paper Company in FY 1982 were 344 billion yen, or roughly 1.4 billion dollars. In all, nine Japanese companies were listed in Pulp & Paper International's 1982 top 100 international pulp and paper firms. In addition to absolute size by sales, most Japanese paper companies have a close link with a large trading company and a bank-oriented industrial grouping. An example of these links and group affiliation for Oji Paper Company, part of the Mitsui Group, is shown in Exhibit 3-31.

Integration

A distinguishing feature of the Japanese industry is the lack of backward integration in most companies. Only one, Oji Paper, owns sufficient timberlands to make a significant contribution to the fiber needs of the company. For the most part, Japanese paper companies either buy fiber on the domestic market or buy it on the international market, mostly from Canada, the US, the USSR, Southeast Asia, and New Zealand. The rest of the needed fiber is converted from wastepaper, which is a more important fiber source for Japan than for any other major paper producer. Pulping of recycled fiber is less energy intensive and reduces the use of expensive virgin fiber, although recycled fiber lowers the quality of the finished product and is inappropriate or unacceptable in many quality-conscious Japanese markets.

Exhibits 3-32 to 3-34 show the increased dependence of the Japanese industry on imported fiber supplies over time. Note that the level of imports has flattened or declined since "chipshock" of 1979-80. The increase in the price of imported chips gave new impetus to development of previously economically unfeasible fiber sources in Japan. "As long as the domestic pulpwoods are supplied at the lower prices than imported counterparts, the share of overseas pulpwoods continues to decline, although the figure is negligible."[11]

The Japanese pulp and paper industry, as a result of dependence on foreign fiber suppliers, has increased forward integration to more converted products such as boxes, packages, and other finished goods. Also evident is diversification to more specialized papers and moves out of paper altogether, into related lines of packaging and containers.

COSTS

The United States

Pulp and paper production costs are difficult to analyze because of the integrated nature of the pulp and paper industry. Outputs of the primary products segments (pulp, paper, and board mills) serve as the inputs to each other and to downstream sectors of the industry. But the industry is one clearly characterized by high capital costs.

Fixed Costs

In a capital-intensive industry such as SIC 26 all participants face high fixed costs. But looking at assets per employee as one measure of capital intensity (refer to Exhibit 3-26), it can be seen that the basic function, pulping, is much more capital intensive than the paper and paperboard manufacturing functions. A cost breakdown of the capital outlay required for each part of the papermaking process shows that 65 percent of the capital cost in an integrated mill goes to the pulping process. Although the costs in Exhibit 3-35 are from 1978 and are now much higher, the percentages are still valid. A new integrated paper mill would cost from $250 to $400 million and up today.

As one moves away from the raw material toward higher value added products, labor becomes relatively more important and fiber and capital become a smaller part of total costs. The importance of fiber and capital in the cost structure of pulp and commodity papers suggests one reason that the expansions in large commodity producing plants have been undertaken primarily by companies with owned fiber supply and sufficient size to provide the financial backing required. It also suggests why the major expansion in the US has been in the South, which has the cheapest per-unit supply of fiber. However, converters and specialty paper producers have remained competitive in the North Central region in spite of older mills and smaller scale because of their stable labor force and positioning with respect to markets and because fiber costs and economies of scale are less important in these higher value added products.

Since high fixed costs, a product of the high cost of capital and the increasingly large scale of pulp and paper mills, are common to all producers, it is the low variable cost producer that will gain market share. "Market share" is extremely important in this industry because of the close and critical connection between operating rates and profitability. It is the low variable cost producer whose operating rates can remain high when demand drops or when overcapacity threatens the industry.

Exhibit 3-36 shows a production cost breakdown for a hypothetical Southern softwood pulp producer, exporting pulp to Europe. The most significant trend to note here is the increase in fixed costs, from 5 percent of total pulp production costs in 1976 to 11 percent in 1982. This increase is due to high construction costs and high interest rates, resulting in higher depreciation charges. This increase is further illustrated in Exhibit 3-37, in which cost of one ton per day (tpd) capacity is projected to increase nine times between 1966 and 1986. The biggest spurts in costs begin in 1978, possibly as a result of high inflation levels, and continue through the early 1980's due to unprecedented high rates of interest. In Exhibit 3-36, we can look at costs in terms of percentages of the total. Fixed costs in 1982 and beyond become a larger cost factor than fiber, which is the primary variable material cost. This suggests that at some point high variable cost older mills with much lower depreciation charges (pre-1976) may become competitive with newer mills which are more efficient and have much lower variable costs, but are carrying very high depreciation costs.

Variable Costs

Variable costs are still primary competitive factors in the pulp and paper industry and as such play a key role in determining the location and nature of new capacity. Variable costs account for 70 to 80 percent of total manufacturing costs in pulp and paper. This compares with variable costs below 50 percent for the chemicals industry and levels near 40 percent for all manufacturing. Capital intensity combined with such high variable costs implies that price competition can be very hard on all but the low-cost producer.

In times of overcapacity or declining demand, the low variable cost producers may in fact be the only survivors. Paper manufacturing operations only break even operating at 90 percent of capacity, based on a 24-hour day and a 360-day year. When demand is slack and orders are scarce, producers still have to run their mills at full capacity because high fixed costs by definition continue even when the mill is not operating. Additionally, the continuous nature of the papermaking process requires continual throughput. Startup and shutdown inefficiencies are enormous. So production levels do not vary much, even in times when capacity exceeds demand.

When demand slackens, the variable that responds is price. As industry operating levels drop toward 90 percent, producers cut prices to stimulate orders. Exhibit 3-28 showed the relative stability of the index of industrial production in pulp and paper compared to other industrial sectors, while Exhibit 3-38 shows the resultant erratic behavior of price indices. These prices are affected by large long-term supply contracts in many commodities such as newsprint; spot market prices are probably even more volatile.

When operating rates are dropping and competition is very keen, it is likely that the lowest cost producer determines price in the market, or the floor below which prices will not fall. Producers with higher variable costs face the dilemma of negative margins or lower operating rates, possibly shutdown. Over the long term a mill with high fixed costs must make enough contribution above variable costs to cover interest and depreciation. But in the short run, older mills with high variable costs can become marginal competitors. This is the situation in the US east coast markets wehre new Southern newsprint and kraft mills are outcompeting older Canadian mills with higher fiber and transportation costs.

Variable costs differ from producer to producer, according to location, degree of integration, and type of product. In the commodity and lower value added products such as pulp, kraft paper, or newsprint, the capital intensity is higher than in the more specialized grades and converted products. This was shown in Exhibit 3-26: Assets per employee are twice to three times as large in pulp mills as they are in paper and paperboard mills. The reverse is also true, and less capital intensity in paper and board implies a greater labor intensity, particularly in the specialty grades. Exhibit 3-39 highlights the differences between various sectors in terms of cost structure.

Fiber Costs

The largest component of paper manufacturing cost is wood fiber. Fiber costs increased 145 percent between 1970 and 1978 although increases have moderated somewhat in recent years. Even in the South, where fiber costs are among the lowest in the world, fiber constituted about 27 percent of total cost of pulp manufacture in 1976, dropping to 23 percent, by 1982.

It would seem simple, then, to determine the lowest-cost regions by looking at local delivered prices of pulpwood. However, the cost of pulpwood is only a proxy for fiber cost. There are many factors which enter into the equation and make it a much more difficult problem than it first appears.

Among the factors influencing fibers costs are flatness of terrain, density of forests, stumpage prices, forest practice regulations, speed of forest rotation, and logging labor costs. Because the southern US has an advantage in virtually all of these factors compared to other regions in the US, it is easy to understand the recent southward movement of the US industry, why it is likely to continue, and its future impact on US competitiveness and pulp-costs decline.

Another effect of fiber cost is the type of product which is most likely to be produced. When fiber costs are low relative to other regions, it is most advantageous to produce products for which virgin fiber is the greatest cost advantage. This is exactly what is happening in the South today. When fiber costs are higher than competing regions, it pays to produce products which rely more on cheaper recycled fiber, as paperboard producers in Japan are doing. While American producers overall use less than 20 percent recycled fiber (wastepaper) in paper production, the figure in Japan is between 45 and 50 percent, and rising. Another tactic which reduces a fiber cost disadvantage is to make higher value added products, such as specialty papers and converted products. Both Japan and Scandinavia have cut back recently on pulping and commodity paper production facilities and are moving to converting and higher-grade paper production.

The correlation between fiber cost and capacity expansion is not perfect, or the cost ranking and the capacity expansion ranking by region would be the same. Obviously other factors enter into the decision to build capacity. As suggested previously, these factors include transportation costs, ownership patterns, infrastructure condition and completeness, distance from market, and in the case of Canadian producers, presence or absence of tariff protection. Other variable cost inputs to pulp and paper production also play an important role.

Labor Costs

On average labor is the second largest cost to US pulp and paper manufacturers after fiber. But the paper production cycle, though more labor intensive than the pulping process, is not especially labor intensive compared to most manufacturing activity. Manufacturing wages per ton of output are only 13-18 percent of the total mill cost for kraft linerboard. This low labor intensity is more marked in high tonnage, high volume production facilities. This can be confirmed by reference to Exhibit 3-39.

The US labor force in paper manufacturing tends to be stable and is very highly unionized. In 1968, 97 percent of US paper workers and 100 percent of pulp workers were unionized. In this organized work force, collective bargaining has been conducted successfully on a mill-by-mill basis. Strikes are a particular threat in an industry where the cost of idled mills is so high and operating rates are so critical to profits.

Lower labor costs in the South are often credited with providing the impetus for the movement of the forest products industry to the South from the Northwest. Overall this could be true, as we see from Exhibit 3-40, which details average hourly earnings of manufacturing production workers by state in the largest paper-producing states.

However, the evidence is less convincing in the case of the pulp and paper industry, SIC 26. In 1923 Southern wage rates in this industry were indeed half those in other areas. By 1967, however, the South had the second highest wage rates in the US industry, and it was second to the Pacific Northwest, the highest region, by only 9 percent. Today paper workers in the South earn around $10 per hour, which is near the national average. Loggers, which are a far less unionized group, work for about $2.50 per hour less than their northern counterparts. It seems clear that the high degree of union organization in the pulp and paper industry has applied upward pressure to wage rates. Exhibit 3-41 shows comparable indices of wage rates for production workers in several segments of the pulp and paper industry, compared to production workers in all manufacturing. While workers in SIC 26 and 2631, "paper and allied products" and "paperboard mills," make historically more than those in general manufacturing, workers in converting plants such as container and box manufacture average about the same as general manufacture. Even so, SIC 26 categories are more highly paid than is common in nondurables manufacture. SIC 26 and 2631 also have seen wages increase more over time than SIC 265, which has increased at about the same rate as all manufacturing.

Wages also increase as a function of increased productivity. Generally speaking, higher productivity workers are paid a higher wage rate, and higher capital intensity usually implies greater productivity per employee. Because of the huge scale of operations in pulp and paper and the high degree of mechanization, output per employee is high. This trend is increasing; as mills increase in size and scale, so does output per worker increase. By union agreement, the wider and faster the machine, the higher the hourly rate of the workers who operate it. In such a unionized industry, it would be hard to justify Southern workers' receiving less than similar workers in other regions, particularly when their mills are the most efficient, high production facilities in the world. Thus, the US appears to have avoided the problems the Canadian industry faces with its east-west wage differentials and resultant labor strife.

We can see from Exhibit 3-42 that an index of wages per manhour in SIC 26 is rising much faster than productivity. Part of this is undoubtedly due to inflation and COLA wage increases, but it suggests that nations with lower wage costs can partially offset the lower fiber cost advantage the South enjoys. Some changes may be occurring in the unbroken escalation of wages, as evidenced by the inability of the Western pulp and paper workers to obtain a shorter contract term in 1978, despite a four-month strike at major Western mills. Concerted resistance by a coalition of paper companies, who continued to operate the mills in the interim using salaried employees, finally caused the unions to capitulate and accept a three-year contract, assuring more operating stability for the mills. Further changes have been occurring in the areas of work rules, flexible pay incentive programs, and other measures designed to increase productivity without increased cost.

Increased productivity is a concern, when one looks at international productivity figures. Between 1960 and 1980, Japanese manufacturing labor productivity increased more than four times, while the US and the UK barely doubled theirs.[12] These are only averages for all manufacturing, but in the pulp and paper industry, German labor productivity increased 30 percent between 1976 and 1980, while the US industry achieved only a 10.4

percent increase over the same period.

Exhibit 3-43 shows pulp and paper employees by country both in absolute numbers and as a percentage of total manufacturing employment.

Energy Costs

The pulp and paper industry is among the top five users of energy in the US. Only primary metals production, chemicals and allied product manufacture, and petroleum refining are significantly greater in terms of energy use. One ton of paper consumes about six barrels of oil in its production, which is more than is required to produce a ton of steel. Here are some other comparisons:

1 pound of aluminum	requires	100,000 Btu's
1 pound of polyethylene	requires	25,000 Btu's
1 pound of paper	requires	15,000 Btu's
1 pound of steel	requires	15,000 Btu's

Strange, 1978

Exhibits 3-44 and 3-45 depict the changing sources of energy in the US pulp and paper industry. As the cost of purchased energy rose 550 percent between 1970 and 1978, the availability and price of energy became another differentiating factor between countries and regions within countries. The cost of electricity is relatively low in the Pacific Northwest due to abundant hydroelectric power generation. This advantage is offset by regional abundance of gas and oil in the South. Total energy costs were estimated at 14 percent of total sales in 1982, on average. As the costs of energy rose, steps were taken to control and reduce energy usage and to increase cogeneration and internal generation of power. The pulp and paper industry leads US industry in production of electric power for its own use. Some producers even own their own hydroelectric generating facilities. Virtually all paper mills now practice some form of cogeneration.[13] Between 35 and 45 percent of all energy used by the industry is self-generated, and most of this energy goes into the pulping process. Besides the immediate cost benefits of cogeneration other economic benefits accrue to the mills by a recently enacted federal statute, the Public Utilities Regulatory Policies Act (PURPA).

Although the papermaking process uses about half of the energy required by the industry, most of this energy is purchased energy. The major need for energy in papermaking is to dry the wet felt and to effectively "iron" the paper on huge heated rolls. Exhibit 3-46 shows the energy use breakdown for the pulp and paper industry.

Exhibit 3-47 graphs the decline in purchased energy and increase in industry energy self-sufficiency. Also an improvement can be seen from the 32.4 million Btu's required in 1972 to the 29.1 million Btu's in 1982.

Between 1972 and 1981 as costs rose, the amount of purchased power consumed per unit of output decreased by more than 30 percent and amount of oil consumed decreased by 43 percent, as a result of more cogeneration, redesign, and general economizing. At the same time total paper production rose more than 17 percent. The net effect of this was that the cost of energy as a percentage of unit sale price doubled, reflected in increased prices for all items. Costs might have been held down more effectively but for increased power consumption due to pollution abatement requirements enacted at the same time as the oil price increases. Most energy used for pollution control is nonproductive; there is no resultant increase in output, and there may be an actual decrease. Another tradeoff concerns cogeneration. For every ton of sludge burned for cogeneration, the amount of fiber recovery per unit of wood decreases. As fiber recovery increases, material available for cogeneration is reduced.

Although there seems to be a compensating energy cost disadvantage between producing regions in the US, all North American regions are better off than paper producers in Japan and Western Europe, who have to import a far greater percentage of their energy. Cost control has helped these industries reduce their energy usage, but rising energy prices continue to accentuate the North American advantage.

Chemical Costs

The pulp and paper industry is the sixth largest industrial user of chemicals in the US. Only the chemicals, textiles, rubber and plastics, agriculture, and petroleum refining industries use more. Each ton of paper or paperboard produced requires 1/4 ton of chemicals.

The largest quantities of chemicals used are process chemicals. Basic inorganics such as salt cake, sulfur, and soda ash are used for pulping, while others such as chlorine and caustic soda are used for bleaching. By far the largest tonnage of any single chemical is kaolin, a fine clay used to impart whiteness, body, opacity, and a smooth finish to paper. Other mineral pigments and fillers include calcium carbonate and talc as fillers, and smaller quantities of titanium dioxide or zinc dioxide as whiteners and opaquers. Additives include synthetic latexes, amino resins, rosin, waxes, dyes, and gums. Paper and pulp manufacture is also the largest user of starches after the food industry.

The complex pulping and papermaking processes require specialty process chemicals to allow increased speed and to make the end product uniform. They include coagulants, defoamers, dispersants, flocculants, humectants, slimicides, flow modifiers, surfactants, and a whole laboratory full of others. These are so specialized that they are often proprietary, in contrast to the commodity chemicals above.

Although chemical costs are substantial, they do not vary much between producers of the same product and process. As such, they do not play a competitive role in the cost of paper products except for some economies of scale. However, chemical costs do vary internationally and hence can affect the relative cost structure of different national industries. Generally, US chemical prices are lower than in most countries due to the strong international competitiveness of the US chemical industry.

Research and Development Costs

Although US pulp and paper firms are large enough to support and benefit from R&D expenditures, it is not a large priority for most producers. The US pulp and paper industry ranked 11th of the 16 basic industry groups in R&D expenditures as a percent of sales in 1982. What R&D there is tends to be very process oriented, and this is usually carried out by chemical and equipment manufacturers. End product research is generally done by end product manufacturers. The exceptions are paper companies involved in the production and sale of consumer products, especially tissue and sanitary products. These companies, such as Kimberly Clark, commonly spend significantly more than the average pulp and paper manufacturer on research and product development.

Exhibit 3-48, taken from Business Week (3/21/84), shows a comparison of pulp and paper industry R&D expenditures compared with the chemicals industry and with all US manufacturing. Exhibit 3-49 displays the change in R&D for the US paper industry over time.

Pollution Abatement Costs

The paper industry is far and away the largest industrial user of water. There has always been a problem supplying the huge quantities of clean water which are a necessity for papermaking. For this reason, pulp and paper mills are always located along waterways, rivers, or lakes. It is not possible to make paper using current technology without a great amount of water for processing and manufacturing. When pulp enters the wet end of a paper machine it contains from 100 to 500 pounds of water for every pound of fiber. The low concentration of this slurry is necessary to assure uniformity of thickness as the paper web is formed; removal of all this water is the chief reason for the gargantuan size of paper machinery (a large paper machine will be over 600 feet long) and much of the energy consumption of the industry.

The papermaking process does not consume all this water but cycles it back to the source. Increasingly with environmental awareness, as well as with the use of more and more expensive chemicals which can be recovered and reused, paper manufacturers need to remove organic and inorganic sludge, chemicals, and process heat before discharging the water.

The progressively stricter federal and state clean water standards have required industry to increase capital expenditures significantly for pollution abatement facilities, with no corresponding increase in productivity as a result. Capital and operating costs for pollution abatement are estimated today at over $15 per ton of paper produced. Although the industry is presently operating in compliance with clean air and water regulations, increased stringency in future standards will further increase costs. The net result is a lowered ability to compete in world markets with producers from countries with less stringent regulatory frameworks. Exhibit 3-50 shows a time series of capital expenditures by the US paper industry on environmental protection, both as a percent of sales and as a percent of total capital expenditures.

The industry has reduced its overall use of processing water per ton of output by 60 percent between 1959 and 1979. In addition to reducing pollution problems and cost of discharge purification, this reduces energy required to remove water from the furnish as well. Another new approach is research directed toward a self-contained paper mill which would internally recycle all the water it uses.

As mentioned elsewhere, air pollution from pulp and paper facilities will remain an odor problem as long as sulfur remains the base for the pulping process. Even a very large improvement in the amount of stack emissions is barely perceptible in reduced odor, and extremely small amounts of dimethyl sulfide (a few parts per billion) can be detected by the human nose. Although great improvements have been made in reduction of stack gases and particulates, current clean air standards completely preclude building of new pulp or paper mills in many areas near urban or pristine "airsheds," and it remains to be seen whether popular and political concern about acid rain will affect the regulation of pulp and paper mills.

Freight Costs

Freight costs are a strong competitive factor in this industry. The product is relatively high-bulk and low-value, and weight-determined transport costs are significant. As a rule, pulp and paper mills are built as close as possible to the raw material source (the forest) to reduce freight costs as far as possible. Converting plants, more labor intensive and market oriented, tend to be located closer to urban high-volume markets. The distance from the resource to the market, the type of intervening transportation, and the rate regulation on that type of transportation are all significant factors in determining market competitiveness. One reason the South is taking market share from Canada in East Coast newsprint markets is that the South's freight and transportation to East Coast cities are better and cheaper. Since few Eastern Canadian mills are built on tidewater, some Southern mills also have a cost advantage to European markets, where Southern pulp, pulpwood, and kraft linerboard producers are cutting inroads into traditional Canadian market share.

Similarly, more of the US Pacific Northwest's production, both of chips and of paper, is exported than the South's because the coastal location of Northwest producers and distance from East Coast markets dictate a cost advantage to the Pacific Rim and an overwhelming cost disadvantage to the East.

Deregulation and differential rail rates have had an impact in US markets as well, favoring Southern producers over the West in most domestic markets.

Japan

Cost Structure

Because of the relatively low level of backward integration, most Japanese firms face a very different cost structure than do most US or Canadian companies. In 1979, Oji Paper's raw material (wood chip) costs were 29 percent of sales, while by comparison fuel oil costs were only 7 percent of sales.

With its cost structure heavily weighted by fiber costs and energy costs compared to the US and Canada, Japan is forced to make changes in its industry strategy: capacity reduction, modernization, rationalization, upgraded value added, and specialization. Research efforts have been carried out entirely by individual companies, both machinery and equipment firms, and paper manufacturing firms. Recently, the Structural Improvements Law (SIL) has facilitated the cooperation of firms for the purpose of R&D.

Previously, the average R&D spending for the industry was about 0.4 percent of gross income, even lower than the US's 1 percent. This was especially low compared to 10 percent for electronics, for example, and 1.61 percent for all Japanese manufacturing. This small R&D emphasis is blamed on the extremely low margins in paper. Stress in existing R&D is placed on securing stable materials supplies, cost reduction, and antipollution measures.

With the SIL, 18 paper companies and nine machinery companies have funded the Paper Technology Research Association--similar to TAPPI in the US. It seems odd that the second largest paper producer in the world had no central trade association to do research before this time. This organization has applied to MITI for subsidies to research higher-yield pulping as well as more use of byproducts. The subsidy would provide 60 percent of costs. Other research topics might include process cost reduction and product development with a market orientation. Promising areas include specialty office papers, papers to meet exacting quality requirements, and technical applications of paper technology to other (nonpaper) fields. However, there was one other group which has done R&D. In 1970 the Japan Pulp and Paper Research Institute was jointly funded by the members of the Oji group primarily to look for a pollution-free pulping technology.

The short-term results of the structural shakedown in the Japanese industry, coupled with the provisions of the Structural Improvements Law, have been an increase in imports of commodity papers into Japan and a simultaneous increase in Japanese paper exports to China, Hong Kong, and Taiwan.

Canada

Costs

Besides productivity concerns, Canada's pulp and paper industry is plagued with instability of labor and frequent work stoppages. Wages paid to militant paper workers in British Columbia are higher than those paid to Eastern Canada counterparts and are higher than those of US workers as well. There is a strong movement in Eastern Canada to achieve parity, which would increase that troubled area's competitive disadvan- tage with the US South. British Columbia paper workers in 1981 struck for 6 weeks, finally achieving a 28 percent wage increase over two years, which is twice the rate of increase for US workers. Not until a pro- longed strike in 1984 did Canadian West Coast paper workers encounter the kind of strong resistance from mills that US workers felt in 1978. The 1984 strike idled 14 percent of the world newsprint capacity for over four months and may have had a long-lasting depressant effect on the region's customer base.

CAPACITY: COMPARATIVE HISTORY AND FUTURE PROJECTIONS

World capacity in paper in the 1970's increased by over 60 million metric tons but 20 million metric tons of outdated capacity were scrapped at the same time, for a net gain of 40 million tons. From an estimated 170 million tons of world paper demand in 1982, Jaakko Poyry projects a demand of 240 million tons (70 million tons increase) by 1990 and 280 million tons (~40 million ton gain) by the mid-1990's. This is a 41 percent net world capacity growth from 1982 to 1990 and a 17 percent net growth in capacity from 1990 to 1995.

Exhibit 3-51 shows past growth and projections for future growth in pulp capacity for several key pulp producing regions. It shows firm projections for 1985 capacity as well as longer range projections from two surveys: one from 1976 showing plans for 1977-87 pulp capacity and one from 1980 showing plans for 1981-91 pulp capacity. It is important to be aware that these projections, made ten years in advance, are more likely to project additions to capacity than dismantling of capacity. Also many of the projected expansions may not be built. Although less inaccurate with respect to additions to capacity than net changes, these rough estimates are useful for what they reveal about changing predictions over time and the accuracy of predictions over time.

For example, most projections became more conservative in 1980 than in 1977. Another survey taken today might show even more conservative projections due to the long world recession and depressed economic projections. The strongest growth projections still come from the fiber-rich regions. Perhaps more notable are the continued increases in projected pulping capacity in Japan and the EC, both regions huge net importers of fiber.

Exhibit 3-52 is similar to the previous exhibit and shows past and future growth patterns in paper capacity. Here the predicitons of growth seem to be much stronger and more certain in all regions. The feeling appears to be even more optimistic in 1980 than in 1976. Somewhat surprising is Japan's continued expansion in paper capacity given some significant domestic overcapacity there. In 1980, however, the Japanese Structural Improvement Law had not yet been applied to the paper industry and expansion may have seemed an appropriate strategy.

The other significant thing about this exhibit is that the major paper producing regions of the world are preparing to increase their capacity about 11 percent between 1980 and 1991. This is only a fraction of the over 40 percent increase in demand between 1982 and 1990 forecast by Jaakko Poyry. Instead of 4 percent per year, a modest enough projection, the paper producing regions are implicitly forecasting a little better than a 1 percent growth rate in paper demand. Where will new capacity come from?

Summarizing the data and industry trade journal comments, it appears that in the next 10 years the major increases in capacity will be in the US and Scandinavia, but Scandinavian net increase in capacity of pulp and paper will be near zero. While US capacity expands, US domestic markets will be growing more slowly than much of the rest of the world--especially the Asian NIC's--creating increased pressure to export fiber rich commodities such as pulp, newsprint, kraftliner, and container board. In 1980 to 1982 US newsprint capacity increased 26 percent to 5.3 million tons, which is more than all the gains in the preceding decade. Pressure is on several developed countries to restructure their industry. France has been especially slow to restructure and Japan has been using its recession cartel, Structural Improvement Law, and foreign investment to address the need for restructuring there. UK cutbacks in capacity have been the most sharp: In the decade 1970-1980 about 20 percent of its paper capacity was closed because it was no longer competitive.

Footnotes

[1]Business Week, March 21, 1984, pp. 202-223.

[2]FAO, Projected Pulp and Paper Mills in the World 1977-87, 1980-1990, Rome: FAO, 1978 and 1981.

[3]Robert Gibbens, "World Pulp and Paper II," Financial Times, April 23, 1980.

[4]Joan E. Huber, ed., The Kline Guide to the Paper Industry, 1980, p. 4.

[5]Kline Guide to the Paper Industry, Fourth Edition, 1980, pp. 33-39.

[6]Thomas P. Clephane and Jeanne Carroll, "Timber ownership still a big factor for success in the paper industry," Pulp & Paper, November 1982, p. 132. The Kline Guide to the Paper Industry (1980) reports Federal Trade Commission data showing the paper industry netting $1.55 in sales for every dollar invested in land, building, and equipment at first cost against an average return for all manufacturing of $2.26. The paper industry did have a higher return than some other capital intensive industries:

$1.16 basic chemicals

$1.39 primary metals

$1.29 petroleum refining.

[7]Richard A. Reese, "Fourdrinier rebuilds: Keys are common sense, proven methods," Pulp & Paper, October 1983, p. 133.

[8]Jaakko Poyry, "Industry prospects, problems for next ten years viewed by expert," Pulp & Paper, May, 1982, p. 191.

[9]Jaakko Poyry, "Industry prospects, problems for next ten years viewed by expert," Pulp & Paper, May 1982, p. 189.

[10]Thomas P. Clephane and Jeanne Carroll, "Timber ownership still big factor for success in the paper industry," Pulp & Paper, November 1982, pp. 132-34.

[11]Japan Pulp and Paper, August 1983, p. 23, "Imports of BKP to jump by ten percent, while pulpwoods decline."

[12]Mary Saso and Stuart Kirby, Japanese Industrial Competition to 1990, 1982.

[13]Basically, cogeneration consists of burning organic materials from sludge (left after pulping) and bark and waste to generate high-pressure steam. The high-pressure steam is then used to power electric turbines. The escaping low-pressure steam is then used as process steam in the pulping process. Since the organic materials burned in this process are a byproduct of pulping, it is the pulping portion of the production cycle which benefits.

EXHIBIT 3-1

Paper Consumption & GNP/Capita

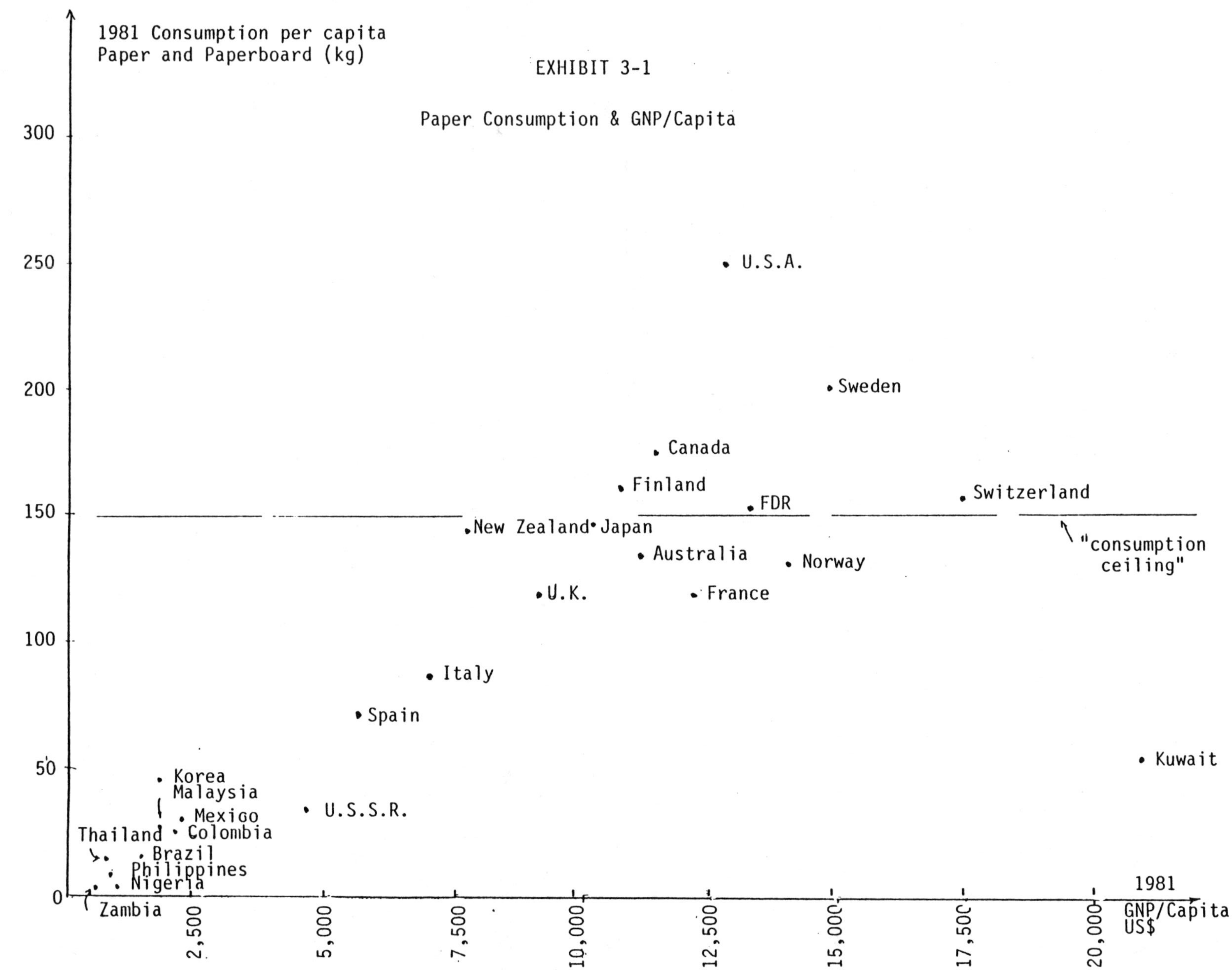

Source: <u>World Bank Atlas</u> and Pulp and Paper International, <u>1983 Annual Review</u>

Distribution of Paper Consumption
in Large Markets
1980

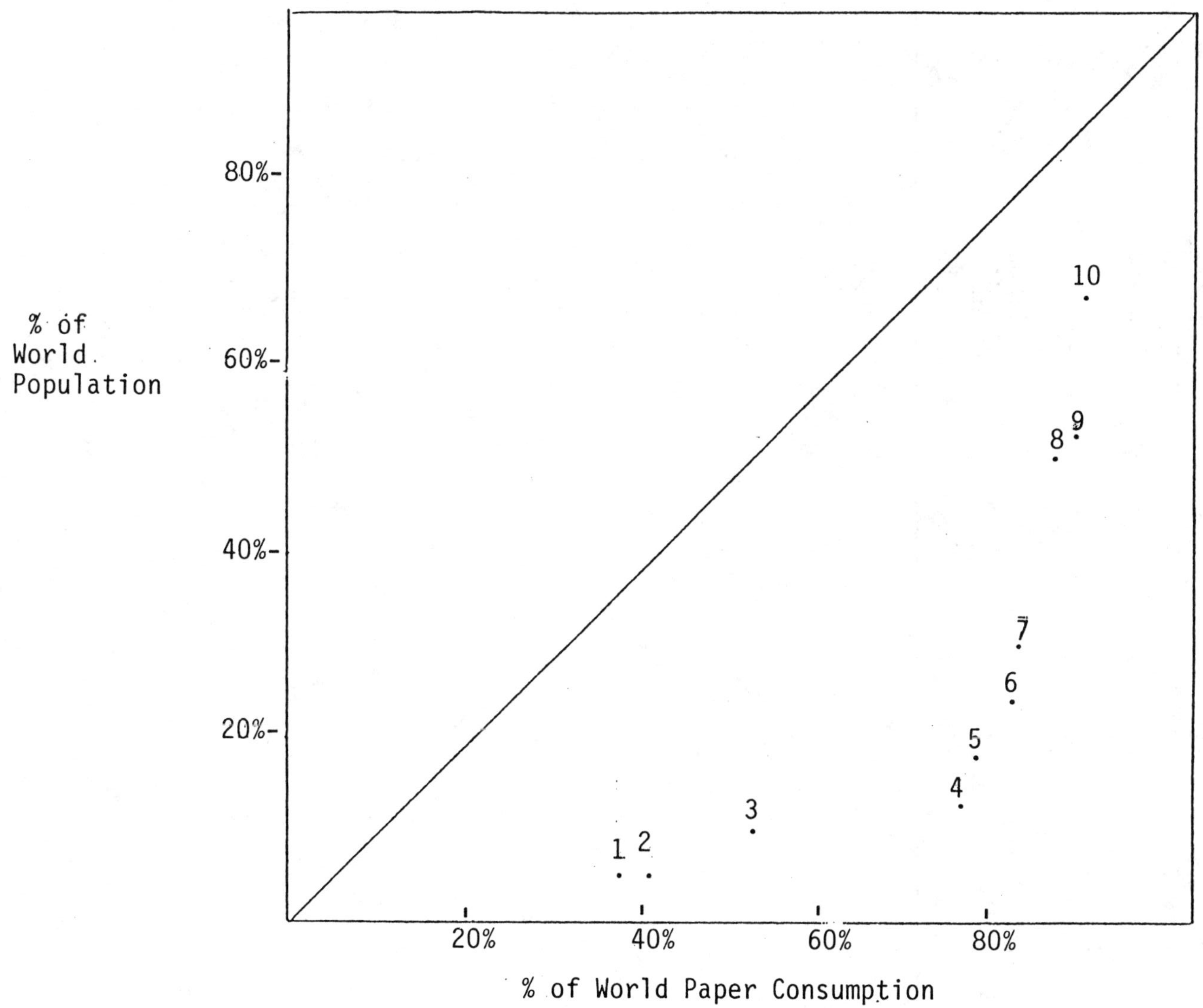

Source: <u>FAO Yearbook of Forest Products</u>, 1980, IBRD World Bank Atlas

Country	Cumulative Population (mns)	Cumulative Population (%)	Cumulative Paper and Paperboard Consumption (1,000 metric tons)	Cumulative Consumption (%)
1. U.S.A.	228	5	63,195	36
2. Canada	252	6	67,290	39
3. Japan	368	8	85,472	49
4. West Europe	738	17	127,880	73
5. Mexico	808	18	130,438	75
6. U.S.S.R.	1,074	25	139,058	80
7. Brazil	1,192	27	142,092	82
8. P.R.C.	2,169	50	151,456	87
9. Indonesia	2,316	53	151,882	87
10. India	2,989	68	153,746	88
Total	4,382	100%	174,315	100%

EXHIBIT 3-2

Paper Consumption & ICP Adjusted GNP/Capita

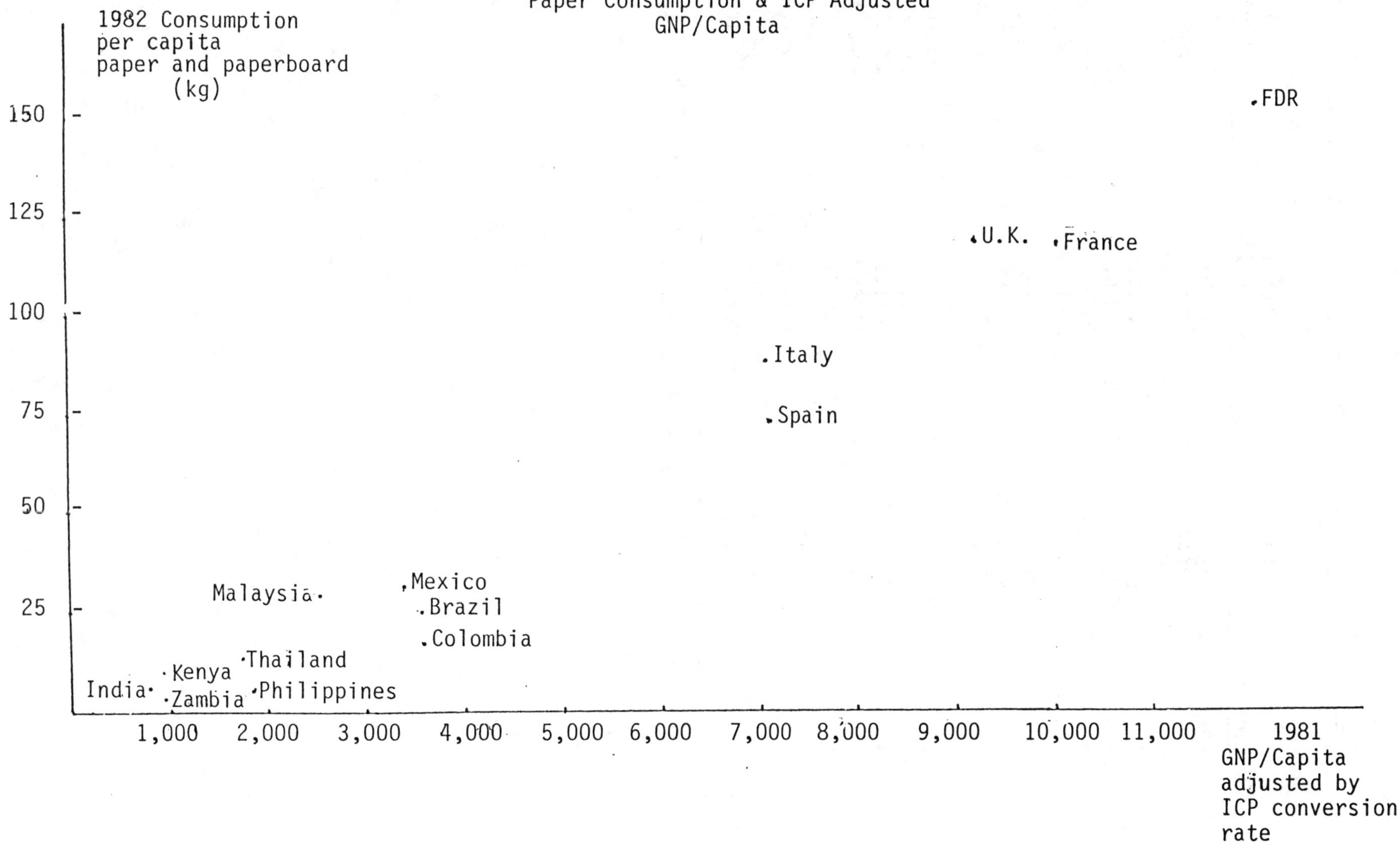

EXHIBIT 3-3

Rank in World	Paper Consumption	Paper Production	Pulp Production
1	U.S.A.	U.S.A.	U.S.A.
2	Japan	Japan	Canada
3	FRG	Canada	U.S.S.R.
4	U.S.S.R.	U.S.S.R.	Japan
5	U.K.	F.R.G.	Sweden
6	France	Sweden	Finland
7	P.R.C.	Finland	P.R.C.
8	Italy	P.R.C.	P.R.C.
9	Canada	France	Brazil
10	Brazil	Italy	France
11	Spain	Brazil	Norway
12	Mexico	U.K.	Spain
13	Netherlands	Spain	Austria
14	Australia	Mexico	South Africa
15	Sweden	Korea	New Zealand
16	Korea	Austria	Czechoslovakia
17	India	Netherlands	Portugal
18	Taiwan	Taiwan	Italy
19	Belgium	Australia	Romania
20	D.R.G.	Norway	Mexico

Source: Pulp and Paper International Annual Review 7/25/83

EXHIBIT 3-4

1982 Paper and Paperboard Consumption				1982 Paper and Paperboard Consumption				1982 Pulp Production			
Rank in World	Country	,000m tons	% of World Consumption		Rank in World	,000m tons	% of World Consumption		Rank in World	,000m ton	% of World Consumption
1	U.S.A.	52,696	35.0	U.S.A.	1	44,212	32.5%	U.S.A.	1	40,715	35.8
2	Japan	15,768	10.5	Japan	2	15,866	10.5	Japan	2	7,843	69.0
3	F.R.G.	8,499	5.6	F.R.G.	5	7,072	4.7	F.R.G.	5	1,822	1.6
4	U.S.S.R.	8,115	5.4	U.S.S.R.	4	8,252	5.4	U.S.S.R.	4	8,327	7.3
5	U.K.	6,073	4.0	U.K.	12	2,840	1.9	U.K.	12	128	0.0
6	France	5,720	3.8	France	9	4,661	3.1	France	9	1,605	1.4
7	P.R.C.	5,364	3.6	P.R.C.	8	5,355	3.5	P.R.C.	8	4,165	3.7
8	Italy	4,440	2.9	Italy	10	4,157	2.7	Italy	10	742	0.7
9	Canada	3,918	2.6	Canada	3	11,278	7.7	Canada	3	15,444	13.6
10	Brazil	2,951	2.0	Brazil	11	2,949	1.9	Brazil	11	2,647	2.3
11	Spain	2,617	1.7	Spain	13	2,440	1.6	Spain	13	1,236	1.1
12	Mexico	2,062	1.4	Mexico	14	1,806	1.2	Mexico	14	680	0.6
13	Netherlands	1,972	1.3	Netherlands	17	1,485	1.0	Netherland	17	124	0.1
14	Australia	1,863	1.2	Australia	19	1,360	0.9	Australia	19	641	0.6
15	Sweden	1,529	1.0	Sweden	6	5,389	3.6	Sweden	6	7,005	0.9
16	Korea	1,509	1.0	Korea	15	1,579	1.0	Korea	15	214	0.2
17	India	1,382	0.9	India	*	1,136	0.8	India	*	n.a.	-
18	Taiwan	1,351	0.9	Taiwan	18	1,415	0.9	Taiwan	18	266**	0.2
19	Belgium	1,284	0.9	Belgium	*	753	0.5	Belgium	*	293	0.3
20	D.R.G.	1,232	0.8	D.R.G.	*	1,146	0.8	D.R.G.	*	616	0.5

% of World Consumption by top 20 86.5%
Total World Consumption 150,696

Top World Production 151,380
*Not in top 20

Total World Production 113,695
** market pulp only

Source: Pulp and Paper International, Annual Review, 7/25/83

EXHIBIT 3-5

1982 Paper and Paperboard Consumption				1982 Paper and Paperboard Consumption				1982 Pulp Production			
Rank in World	Country	,000m tons	% of World Consumption	Country	Rank in World	,000m tons	% of World Consumption	Country	Rank in World	,000m ton	% of World Consumption
1	U.S.A.	52,696	35.0	U.S.A.	1	49,212	32.5	U.S.A.	1	40,715	35.8
2	Japan	15,768	10.5	Japan	2	15,866	10.5	Canada	2	15,444	13.6
3	F.R.G.	8,499	5.6	Canada	3	11,278	7.5	U.S.S.R.	3	8,327	7.3
4	U.S.S.R.	8,115	5.4	U.S.S.R.	4	8,252	5.5	Japan	4	7,843	6.9
5	U.K.	6,073	4.0	F.R.G.	5	7,072	4.7	Sweden	5	7,005	6.2
6	France	5,720	3.8	Sweden	6	5,389	3.6	Finland	6	6,104	5.4
7	P.R.C.	5,364	3.6	Finland	7	5,359	3.5	P.R.C.	7	4,165	3.7
8	Italy	4,440	2.9	P.R.C.	8	5,355	3.5	Brazil	8	2,647	2.3
9	Canada	3,918	2.6	France	9	4,661	3.1	F.R.G.	9	1,822	1.6
10	Brazil	2,951	2.0	Italy	10	4,157	2.7	France	10	1,605	1.9
11	Spain	2,617	1.7	Brazil	11	2,949	1.9	Norway	11	1,384	1.2
12	Mexico	2,062	1.4	U.K.	12	2,840	1.9	Spain	12	1,236	1.1
13	Netherlands	1,972	1.3	Spain	13	2,440	1.6	Austria	13	1,081	1.0
14	Australia	1,863	1.2	Mexico	14	1,806	1.2	South Africa	14	1,059	0.9
15	Sweden	1,529	1.0	Korea	15	1,579	1.0	New Zealand	15	955	0.8
16	Korea	1,509	1.0	Austria	16	1,552	1.0	Czecho-slovakia	16	873	0.8
17	India	1,382	0.9	Netherland	17	1,485	1.0	Portugal	17	865	0.8
18	Taiwan	1,351	0.9	Taiwan	18	1,415	0.9	Italy	18	742	0.7
19	Belgium	1,284	0.9	Australia	19	1,360	0.9	Romania	19	741	0.7
20	D.R.G.	1,232	0.8	Norway	20	1,185	0.8	Mexico	20	680	0.6
% of World Consumption by top 20 86.5%				% of World Production by top 20 89.3				% of World Production by top 20 92.8			
Total World Consumption 150,696				Total World Production 151,380				Total World Production 113,695			

Source: Annual Review, <u>Pulp and Paper International</u>, 7/25/83

EXHIBIT 3-6

Paper Consumption and Production and Pulp Production in Major Markets

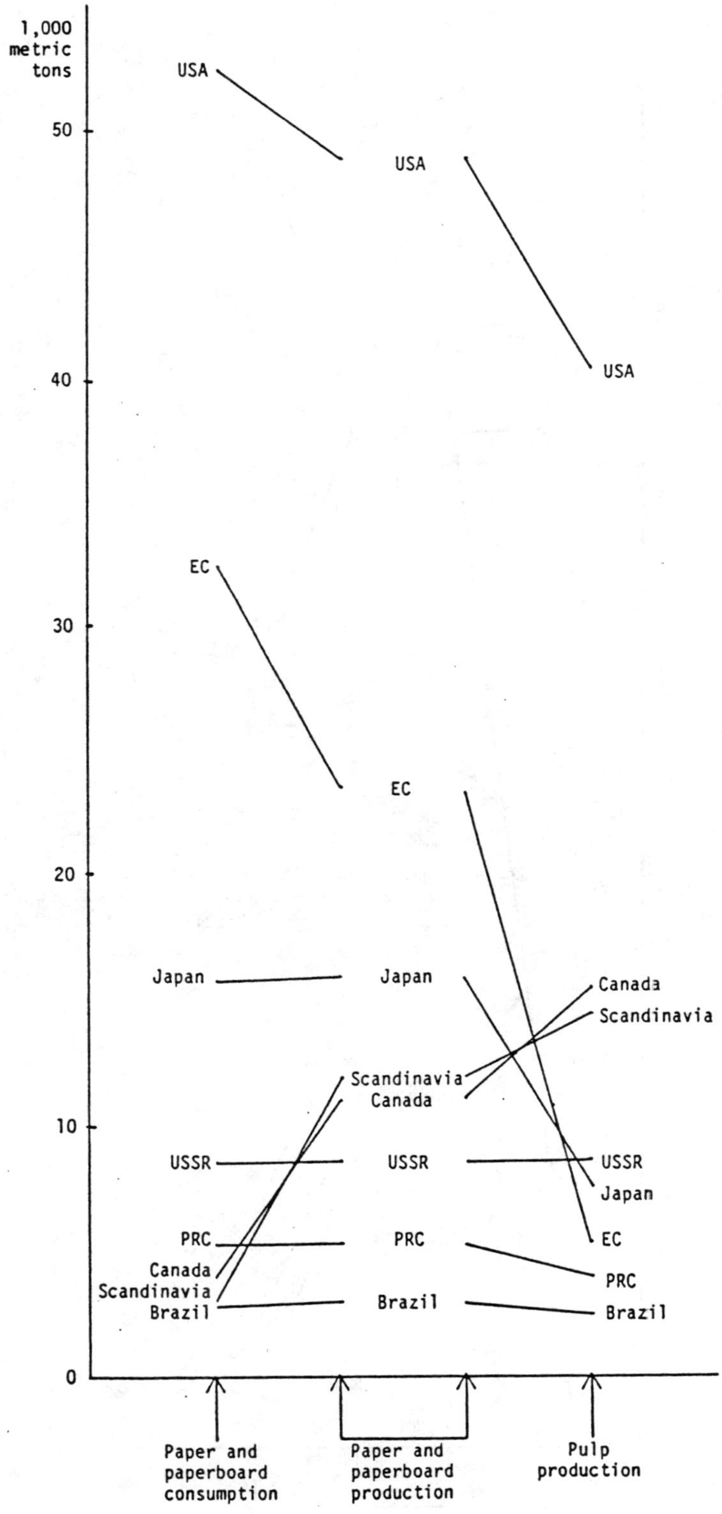

EXHIBIT 3-6 (continued)

Paper Consumption and Production and Pulp Production in Major Markets

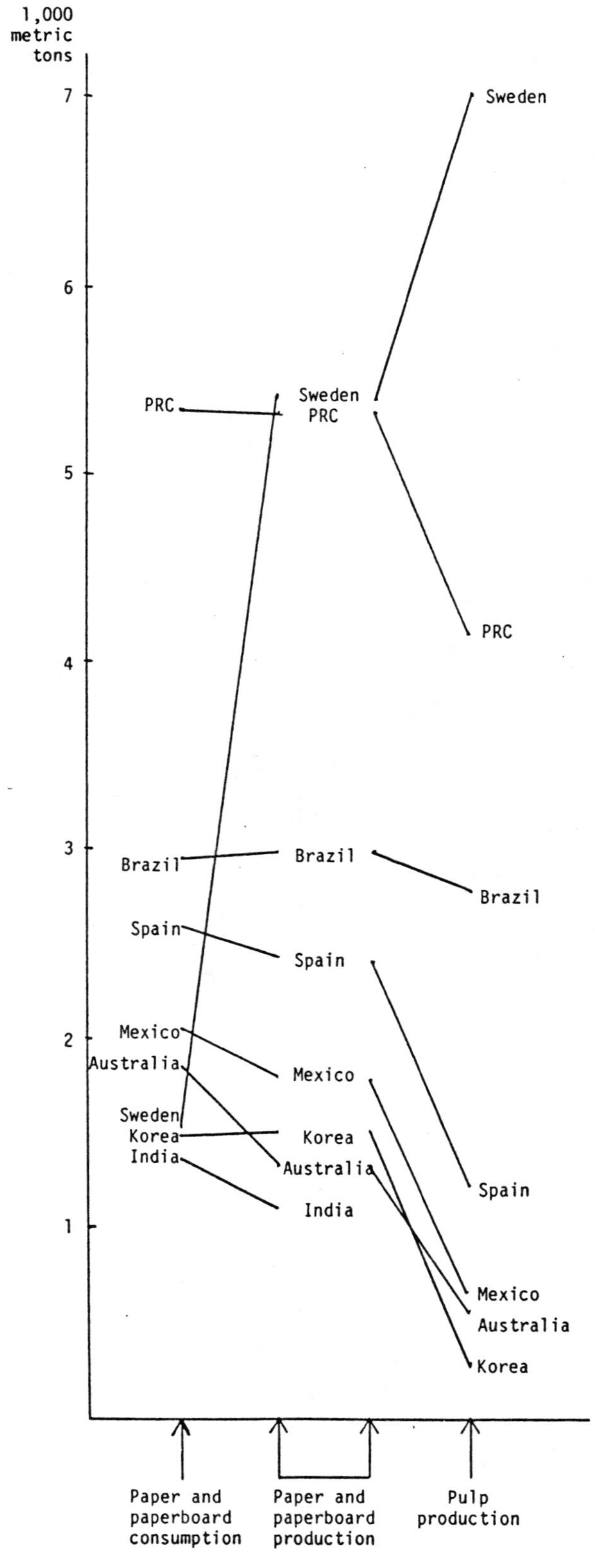

EXHIBIT 3-7

Per Capita Consumption
(kg)

		1982	1981	1975
1.	U.S.A.	250	268	238
2.	Sweden	202	200	195
3.	Canada	175	197	159
4.	Denmark	166	163	134
5.	Finland	160	198	136
6.	Switzerland	156	166	110
7.	F.R.Germany	152	157	112
8.	Netherlands	152	154	124
9.	Japan	146	143	117
10.	Belgium	143	141	102
11.	New Zealand	143	135	123
12.	Australia	135	137	115
13.	Norway	130	141	135
14.	U.K.	119	124	108
15.	France	117	115	89
16.	Austria	104	108	79
17.	Iceland	90	91	--
18.	Ireland	88	93	71
19.	Italy	87	87	--
20.	Taiwan	87	78	--

Source: Pulp and Paper International Annual Review, July 25, 1983.

EXIBIT 3-8

End Use Per Capita Paper and Paperboard Consumption

Imusa 1920-1980

	Paper							Paperboard	Total
	News-print	Groundwood and book	Writing and related	Packaging and converting	Sanitary and tissue	Other	Total paper		
1920	29%	13%	5%	13%	3%	8%	71%	29%	100%
1930	28	14	6	12	3	5	68	32	100
1940	23	13	4	14	4	5	67	33	100
1950	20	11	4	11	5	7	58	42	100
1960	19	12	4	10	6	5	56	44	100
1970	17	13	5	9	6	5	55	45	100
1980	17	24		8	6	4	59	41	100

Source: Guthrie, Economics of Pulp and Paper, 1972, p. 52 and API, 1983 Statistics of Paper, Paperboard and Wood Pulp 8/83.

EXHIBIT 3-9

Growth in Wood Pulp Capacity
(metric tons)

	US	Canada	Japan	Oceania	Scandi-navia	EC
1970 — 1975	+12%	+15%	+30%	+81%	+5%	+10%
1975 — 1980	+11	-1	+6	+8	-1	-1
1980 — 1985^p	+11	+9	-3	+19	+2	+1

p = projected

Source: OECD, The Pulp and Paper Industry, 1983.

EXHIBIT 3-10

Growth in Paper and Board Capacity
(metric tons)

	US	Canada	Japan	Oceania	Scandi-navia	EC
1970 — 1975	+12%	+12%	+41%	+20%	+28%	+13%
1975 — 1980	+9	+3	+9	+22	+9	+5
1980 — 1985^p	+10	+13	+7	+17	+14	+6

p = projected

Source: OECD, The Pulp and Paper Industry, 1983.

EXHIBIT 3-11

US Regional Capacities: Pulp, Paper and Board

	Northeast[1]	North Central	West	South	Total
Pulp					
1970	9%	9%	19%	63%	100%
1980	9	8	17	66	100
1985[p]	8	7	16	68	100
Newsprint					
1970	16	3	26	55	100
1980	na	3	26	60	100
1985[p]	na	na	31	58	100
Total Paper					
1970	27	22	14	37	100
1980	26	21	14	39	100
1985[p]	25	22	14	39	100
Total Paperboard					
1970	11	16	15	58	100
1980	8	14	15	63	100
1985[p]	7	14	15	64	100
Total Paper and Paperboard					
1970	19	20	14	47	100
1980	17	18	14	51	100
1985[p]	16	18	14	52	100

[1]Northeast = New England and Mid Atlantic

p = projected

Source: API, <u>1983 Statistics of Paper, Paperboard and Woodpulp</u>, August 1983.

EXHIBIT 3-12

US Pulp and Paper Industry Capital Spending Plans, 1983-85

State	Rank	Capacity investment	Environmental investment	Total
South Carolina	1	$1,093	$133	$1,226
Alabama	2	1,016	58	1,074
Washington	3	710	65	775
Mississippi	4	556	85	641
Michigan	5	470	77	547

EXHIBIT 3-13

Paper, Manufacturing, and the Canadian Economy

	Value of Shipments ($C millions)	% of Total Mfg.	Value Added ($C millions)	% of Total Mfg.	Total Employees (1,000)	% of Total Mfg.	Salaries & Wages ($C millions)	% of Total Mfg.
Paper and allied industries								
1978	10,197	(7.9%)	4,539	(8.8%)	127	(7.1%)	2,282	(8.6%)
1981	14,502	(8.6)						
1982	15,534	(8.2)						
Wood industries								
1978	7,477	(5.8)	3,507	(6.8)	119	(6.6)	1,821)	(6.9)
1981	8,395	(5.0)						
1982	8,579	(4.5)						
Food and beverages								
1978	21,956	(17.0)	6,921	(13.4)	230	(12.8)	3,181	(12.0)
1981	28,313	(16.8)						
1982	31,580	(16.7)						

Source: Canada Statistical Review (12/81, 6/83) and Canada Yearbook 1980/81

EXHIBIT 3-14

Canadian Exports, 1980

All Canadian production	Share of production exported	Share of production exported to US
Pulp	37%	50%
Newsprint	89	82
Paper and paperboard	71	63

EXHIBIT 3-15

Selected US and Canadian Tariffs, 1984

Product	USTS-A Column 1, mfn rate of duty	Canadian tariff
Woodchips, other than waste	Free	Free
Woodpulp	Free	Free
Newsprint	Free	Free
Groundwood printing paper		
Uncoated	Free	4.7% + 9% Federal Sales Tax
Coated	4.0%	6.3% + 9% Federal Sales Tax
Tissue	6.5%	12.9% + 9% Federal Sales Tax
Wallpaper	1.9%	10.3% + 9% Federal Sales Tax
Writing paper	5.0%	11.4% + 9% Federal Sales Tax
Paper envelopes, plain	4.7%	12.9% + 9% Federal Sales Tax

Source: TSUS-A, 1984 and Customs Canada.

EXHIBIT 3-16

Japan's Paper and Paperboard Supply
and Demand
(1,000 metric tons)

Year	Production	Import	Exports	Apparent Consumption	Per Capita Consumption (inkg)
1960	4,513	1	169	4,345	47
1970	12,973	100	508	12,565	121
1973	15,974	260	465	15,769	145
1975	13,601	107	654	13,054	117
1980	18,088	495	655	17,928	153
1981	16,980	490	673	16,797	142
1983	17,453	632	740	17,345	146

Source: Japan Paper Association, Pulp & Paper Statistics 1983.

EXHIBIT 3-17

End Use Per Capita Paper and Paperboard
Consumption in Japan
1960 - 1982

	Paper				Paperboard
	News-print	Writing, Printing, and Misc.	Packaging	Tissue	
1960	17%	30%	13%	4	37
1982	16%	33%	5%	5	41

Source: Calculated from Japan Paper Association, Pulp and Paper Statistics, 1983.

EXHIBIT 3-18

Japan's Paper Trade 1960-1982

(1,000 metric tons) [% domestic production]

	1960		1970		1975		1980		1982	
IMPORTS										
Newsprint	-	[-]	87.5	[5]	29.4	[1]	126.9	[5]	243.6	[9]
Total paper	.5	[-]	97.7	[1]	86.5	[1]	219.7	[2]	351.3	[3]
Kraft linerboard	-		.4	[-]	13.9	[1]	206.1	[14]	203.1	[16]
Total paperboard	.5	[-]	1.6	[-]	19.8	[-]	273.5	[4]	278.6	[4]
All paper and paperboard	1.0	[-]	99.5	[1]	106.8	[1]	494.7	[3]	631.5	[4]
EXPORTS										
Newsprint	8.9	[1]	31.9	[2]	107.3	[5]	97.5	[4]	121.1	[5]
Printing & writing	97.9	[10]	189.2	[8]	189.7	[7]	127.7	[4]	244.2	[6]
Total paper	140.2	[5]	268.1	[4]	355.9	[5]	307.4	[3]	411.2	[4]
Whiteboard	14.1	[4]	113.1	[11]	101.2	[11]	118.0	[9]	181.0	[14]
Kraft linerboard	1.2	[-]	24.1	[3]	84.2	[7]	125.5	[9]	66.6	[5]
Total paperboard	28.9	[2]	239.6	[4]	297.6	[5]	348.5	[5]	328.6	[4]
All paper and paperboard	169.0	[4]	507.8	[4]	653.6	[5]	655.9	[4]	739.8	[4]

Source: Japan Paper Association, Pulp and Paper Statistics, 1983.

EXHIBIT 3-19

BROAD OUTLINE OF JAPANESE PULP AND PAPER INDUSTRY'S ACTIVITIES IN 1982

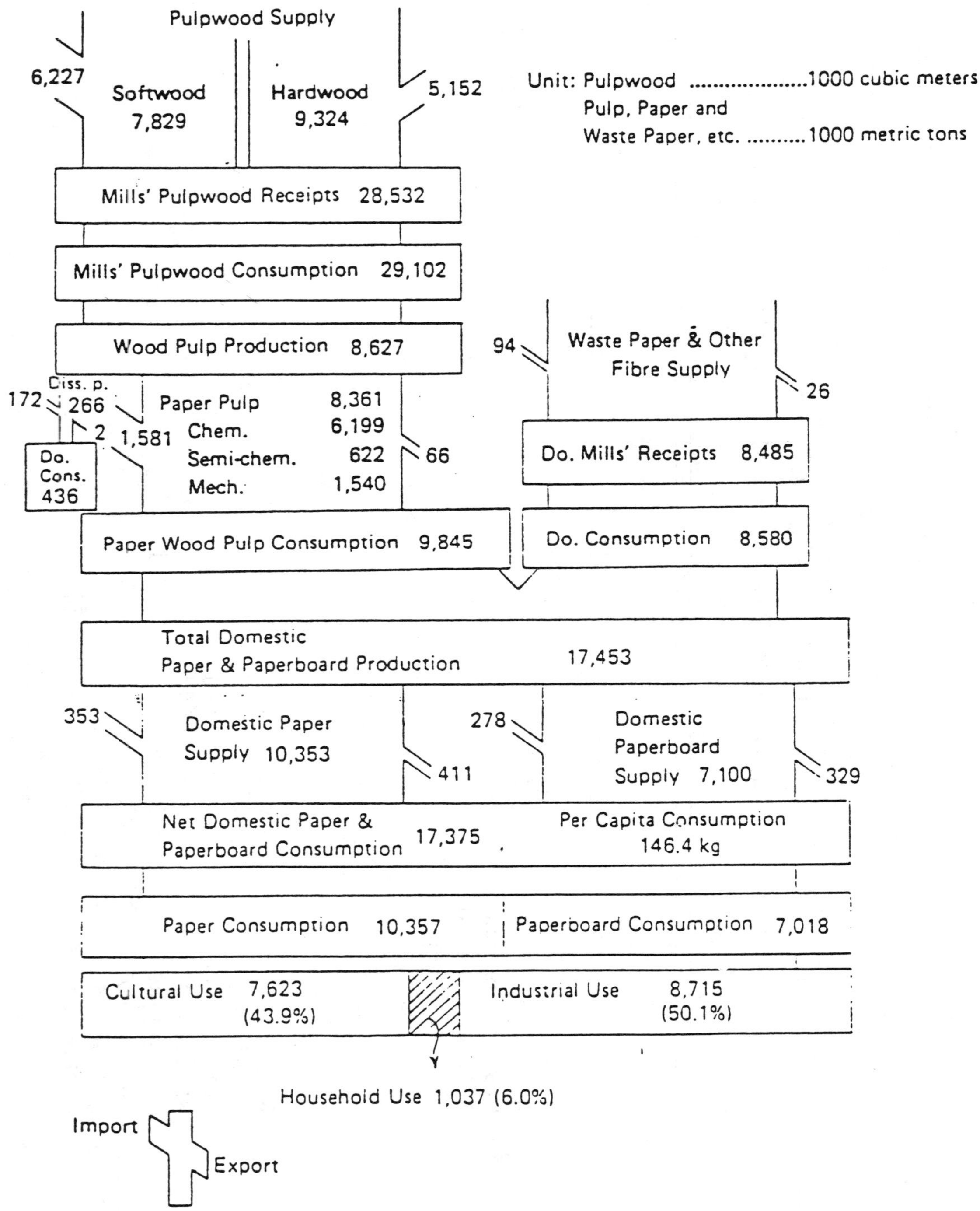

Source: Japan Paper Association, Pulp and Paper Statistics, 1983

EXHIBIT 3-20

Japanese Import Dependence on Pulpwood

	Pulpwood supplied by imports		Source of imported pulpwood				
	1,000m^3	Share of total	US	Australia	Canada	USSR	Other
1960	193	2%				100%	
1965	461	3	55%			38	7%
1970	5,285	19	78			5	17
1975	11,791	41	66	17%		6	11
1980	15,570	44	52	24	8%	4	12
1981	12,321	43	50	26	7	5	12
1982	11,379	40	46	30	6	4	14

Source: Japan Paper Association, Pulp and Paper Statistics, 1983.

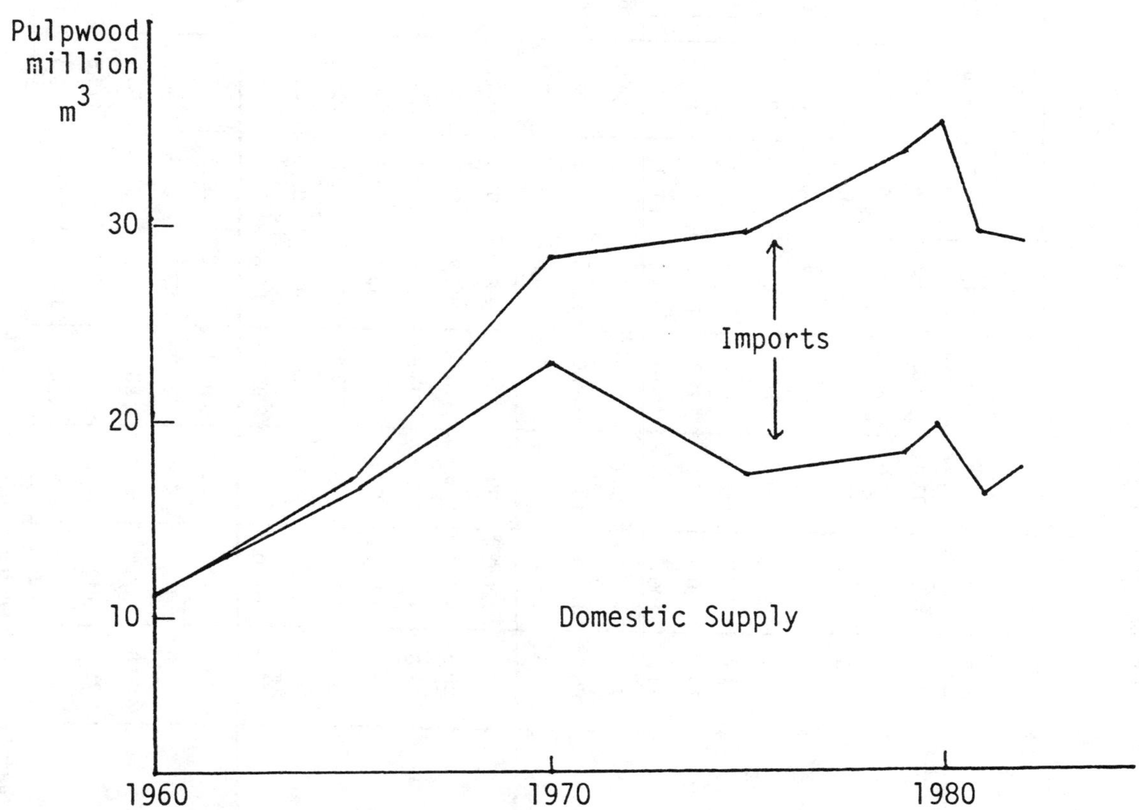

EXHIBIT 3-21

Wholesale Price Indexes of Pulp, Paper and Related Products 1979 ~ 1982

(1980 average = 100)

	1979	1980	1981	1982	1982 J	F	M	A	M	J	J	A	S	O	N	D
All commodities	87.0	100.0	101.4	101.9	101.8	101.8	101.8	101.9	101.7	101.5	101.7	102.1	102.3	102.0	102.1	102.0
Pulp, paper & related products	76.5	100.0	97.0	95.2	96.2	96.3	96.5	96.6	96.3	95.8	94.8	94.2	93.9	93.8	93.8	93.8
Pulp	76.9	100.0	87.0	86.5	86.8	86.4	86.4	86.4	86.8	87.2	87.0	87.0	86.8	86.2	85.8	85.6
Dissolving sulfite pulp	83.5	100.0	90.9	92.1	93.0	93.0	93.0	93.0	93.0	91.4	91.4	91.4	91.4	91.4	91.4	91.4
Bl. sulfate paper pulp	76.1	100.0	86.4	87.5	87.4	86.6	86.6	86.6	87.4	88.5	88.5	88.5	88.1	87.7	87.0	87.0
Paper & paperboard	76.0	100.0	99.2	97.7	99.9	100.0	99.9	99.3	98.5	97.6	96.8	96.1	96.0	96.0	96.3	96.2
Newsprint	84.2	100.0	111.3	111.3	111.3	111.3	111.3	111.3	111.3	111.3	111.3	111.3	111.3	111.3	111.3	111.3
Woodfree printing	79.2	100.0	95.9	101.7	101.9	101.9	101.9	101.9	101.9	101.9	101.9	101.9	101.9	101.7	101.0	100.3
Coated paper	83.0	100.0	94.1	100.2	100.5	100.5	100.5	100.5	100.5	100.4	100.1	100.1	100.1	100.1	99.7	99.4
Kraft paper	75.7	100.0	94.6	97.9	99.5	99.2	98.5	98.5	98.5	98.4	98.5	98.5	97.6	96.2	95.9	95.5
Linerboard	66.5	100.0	101.7	96.5	100.8	100.8	100.8	100.1	98.4	96.7	95.3	94.2	94.0	93.5	92.4	91.5
White board	75.5	100.0	93.7	94.4	94.8	95.3	95.2	94.9	94.9	94.9	94.6	94.2	93.5	93.5	93.5	93.5
Paper products	77.5	100.0	98.0	96.3	96.4	96.6	96.8	97.1	97.2	97.1	96.3	96.0	95.6	95.5	95.5	95.4
Paper supplies for household	88.5	100.0	95.2	98.5	95.2	96.6	97.8	98.2	98.4	98.4	99.6	99.8	100.1	99.5	99.3	99.8
Paper supplies for office & school	81.8	100.0	103.5	103.6	103.5	103.5	103.5	103.5	103.5	103.6	103.6	103.6	103.6	103.6	103.6	103.6
Paper & paperboard containers	75.2	100.0	97.6	94.9	95.6	95.6	95.7	96.0	96.0	96.0	94.7	94.3	93.8	93.8	93.7	93.6
Paper bag	83.4	100.0	100.8	102.6	100.6	100.6	101.3	102.4	103.1	103.4	103.4	103.4	103.4	103.4	103.4	102.8
Corrugated shipping containers	69.8	100.0	96.3	92.3	93.3	93.3	93.4	93.7	93.7	93.7	92.2	91.5	90.8	90.8	90.8	90.6
Carton	92.3	100.0	100.6	98.5	99.0	99.0	99.0	99.0	99.0	99.0	97.9	97.9	97.9	97.9	97.9	97.9
Pulpwood (Hardwood chip)	75.8	100.0	96.8	91.9	93.6	93.6	93.3	92.5	92.3	91.9	91.3	91.3	91.0	90.9	90.9	90.6
Wastepaper (News)	71.9	100.0	6.15	64.0	56.8	56.8	56.8	63.1	63.1	63.1	63.1	63.1	63.1	64.7	69.4	85.2
(Corrugated box)	66.6	100.0	62.8	56.7	53.9	53.9	53.9	53.9	53.9	53.9	53.9	53.9	53.9	58.0	62.0	75.5
Caustic soda	76.1	100.0	111.3	118.5	118.5	118.5	118.5	118.5	118.5	118.5	118.5	118.5	118.5	118.5	118.5	118.5
Heavy fuel oil (C grade)	49.5	100.0	118.0	124.1	123.1	123.0	123.0	123.0	123.0	123.0	123.0	123.0	126.2	126.2	126.2	126.2

Sources: Bank of Japan and Paper Recycling Promotion Center

Source: Japan Paper Association, Pulp and Paper Statistics 1983.

EXHIBIT 3-23

	Fiber Supply[1]				Pulp Production[1]		
	Roundwood Production[2] (1,000 m3)	% of World Removals	Cumulative[3] %		Production 1,000 metric tons	% of World Production	Cumulative[4] %
1. U.S.S.R.*	356,000	11.8	11.8	1. U.S.A.	45,835	36.1	36.1
2. U.S.A.*	322,265	10.7	22.5	2. Canada	19,745	15.6	51.7
3. China*	224,628	7.4	29.9	3. Japan*	9,975	7.9	59.6
4. Brazil*	217,324	7.2	37.1	4. U.S.S.R.*	8,730	6.9	66.5
5. India*	214,671	7.1	44.2	5. Sweden	8,699	6.9	73.4
6. Canada*	161,366	5.3	49.5	6. Finland	7,440	5.9	79.3
7. Indonesia*	157,184	5.2	54.7	7. Brazil*	3,338	2.6	81.9
8. Nigeria	99,484	3.3	58.0	8. China*	2,346	1.9	83.8
9. Korea*	68,437	2.3	60.3	9. F.R.G.	1,996	1.6	85.4
10. Vietnam*	64,566	2.1	62.4	10. France	1,909	1.5	86.9
11. Sweden	53,411	1.8	64.2	11. Norway	1,602	1.3	88.2
12. Finland	47,961	1.6	65.8	12. Spain	1,253	1.0	89.2
				13. Austria	1,228	1.0	90.2
				14. New Zealand*	1,043	0.8	91.0

[1] 1980

[2] Defined as all forest removals

[3] 1980 World fiber estimate FAO = 3,020,306 X 1000 m3

[4] 1980 World Woodpulp production FAO estimate = 126,755 X 1000 metric tons

EXHIBIT 3-24

Concentration in the Paper Industry

	Total		Value of Shipments concentration ratio for 4 largest companies: 100 to 80%		79 to 60%		59 to 40%		39 to 20%		Less than 20%	
	Number of industries in two digit code	Value of shipments ($millions)	Number of industries	Value of shipments	Number of industries	Value of shipments	Number of industries	Value of shipments	Number of industries	Value of shipments	Number of industries	Value of shipments
20 Food and kindred products	47	192,912	2	3,064 (2%)	8	24,695 (12%)	16	34,752 (18%)	17	98,147 (51%)	4	32,204 (17%)
22 Textile mill products	30	40,551	1	1,013 (2)	3	2,658 (7)	9	10,663 (26)	14	19,030 (47)	3	6,883 (17)
24 Lumber and wood products	17	39,919	-	-	-	- -	1	453 (1)	7	17,651 (44)	9	21,814 (55)
26 Paper and Allied products	17	52,086	1	240	1	4,921 (9)	5	6,682 (13)	7	30,833 (59)	3	9,409 (18)
33 Primary metal	26	103,179	4	6,049 (6)	3	11,549 (11)	7	57,296 (55)	11	27,730 (27)	1	553 (1)
37 Transportation equipment	18	166,954	2	77,452 (46)	6	49,629 (30)	6	33,843 (20)	3	4,207 (3)	1	1,822 (1)

Source: Census of Manufactures, 1977, Concentration Ratios in Manufacturing, 9-10.

EXHIBIT 3-25

Value of Industry Shipments

	Number of companies	Percent accounted for by largest 4 firms	8 firms	20 firms	50 firms	Total value of industry shipments ($ million)
2611 Pulp mills						
1977	26	48	76	99	100	2,091.1
1967	46	45	70	97	100	730.5
1963	34	48	72	97	100	609.1
2621 Paper mills						
1977	171	23	42	70	92	12,613.3
1967	203	26	43	65	86	4,844.0
1963	186	26	42	63	85	3,824.9
2631 Paperboard mills						
1977	120	27	42	70	95	7,124.3
1967	148	27	42	67	92	2,907.0
1963	146	27	43	67	90	2,315.2
2641 Paper coating and glazing						
1977	454	30	45	64	82	3,392.0
1967	334	38	49	68	85	1,566.5
1963	323	30	43	64	85	1,161.6
2643 Bags						
1977	462	26	40	58	75	3,482.3
1967	466	23	38	59	76	1,375.6
1963	495	22	36	58	76	1,061.7
2647 Sanitary paper products						
1977	72	65	84	96	99	4,921.2
1967	91	63	79	92	99	1,293.4
1963	97	62	76	90	99	983.0
2649 Converted paper products n.e.c.						
1977	783	17	25	38	58	1,697.5
1972	287	22	32	52	72	630.6
2651 Folding paperboard boxes						
1977	452	22	35	56	74	2,579.0
1967	472	22	36	54	70	1,216.2
1963	534	21	37	55	69	1,081.1
2653 Corrugated and solid boxes						
1977	863	19	33	58	75	7,351.1
1967	568	18	33	61	78	2,959.6
1963	534	20	36	63	79	2,166.1

Source: Census of Manufacturers, 1977, Concentration Ratios in Manufacturing 9-27

EXHIBIT 3-26

Assets Per Employee in Pulp and Paper
Compared to All Manufacturing

	1963	1967	1972	1976
All manufacturing	$ 9,294 (100)	11,268 (100)	15,799 (100)	21,423 (100)
2611 Pulp mills	58,728 (630)	79,576 (710)	14,387 (720)	156,178 (730)
2621 Paper mills	31,902 (340)	42,768 (380)	61,622 (390)	81,060 (380)
2661 Paperboard mills	24,496 (260)	33,513 (300)	30,302 (190)	44,539 (210)

Source: 1977 Census of Manufacturers.

EXHIBIT 3-27

Self Sufficiency And Timber Ownership

Timber consumption and growth data

Major forest products/ paper companies	Timber inventory (Million ft³)	1981 net consumption (Million ft³)	Estimated timber growth	1981 timber harvest	Timber growth to inventory	Ratio of timber growth to consumption	Timber harvest to consumption
Weyerhaeuser*	10,768	762	661	743	6.14	86.75	97.51
International Paper*	8563	782	453	274	5.29	57.93	35.04
Georgia-Pacific*	5706	687	327	350	5.73	47.60	50.95
Crown Zellerbach*	2865	268	162	180	5.65	60.45	67.16
Champion International*	3539	397	224	177	6.33	56.42	44.58
St. Regis Paper*	4068	397	165	130	4.05	41.56	32.75
Boise Cascade*	3731	376	154	172	4.13	40.96	45.74
Pacific Lumber	1336	19	22	19	1.65	115.79	100.00
Scott Paper*	2549	230	101	99	3.96	43.91	43.04
Potlatch	2446	151	78	94	3.19	51.66	62.25
Union Camp	1728	302	156	90	9.03	51.66	29.80
Louisiana-Pacific*	1656	275	73	55	4.41	26.55	20.00
Willamette	1057	146	53	44	5.01	36.30	30.14
Mead*	1816	313	93	69	5.12	29.71	22.04
Great Northern Nekoosa	2871	293	79	95	2.75	26.96	32.42
Longview Fibre	713	82	52	22	7.29	63.41	26.83
Masonite	580	122	48	42	8.28	39.34	34.43
Westvaco*	1350	221	79	77	5.85	35.75	34.84
Kimberly-Clark*	620	102	51	19	8.23	50.00	18.63
Chesapeake Corp. of Va.	291	67	29	9	9.97	43.28	13.43
Federal Paper Board	286	93	34	13	11.89	36.56	13.98
Hammermill Paper	444	131	25	16	5.63	19.08	12.21
Total	**58,983**	**6216**	**3119**	**2789**	**5.29**	**50.18**	**44.87**

*U.S. operations only

Source: Morgan Stanley

Source: Thomas P. Clephane and Jeanne Carroll, "Timber ownership still big factor for success in the paper industry," Pulp and Paper, November, 1982, p. 135.

EXHIBIT 3-28
Index Of Industrial Production
(1967=100)

SOURCE: Federal Reserve Monthly Statistical Releases-Industrial Production

EXHIBIT 3-29

Age of Japanese Facilities

Lapse of Years after Installation and Production Capacities of Existing Pulp and Paper Making Facilities of 50 Main Companies as of Feb., 1983

Lapse of years	Sulfate pulping facilites				Paper machines				Paperboard machines			
	Num.	%	t/d	%	Num.	%	t/d	%	Num.	%	t/d	%
0 ~ 4	9	5.1	1,708	7.1	18	5.1	3,349	11.1	2	1.8	928	4.9
5 ~ 9	42	23.7	6,446	26.7	21	6.0	3,550	11.8	13	11.8	3,981	21.2
10 ~ 14	35	19.8	6,373	26.4	63	17.9	6,925	22.9	25	22.7	5,811	31.0
15 ~ 19	13	7.3	2,067	8.6	56	16.0	4,827	16.0	31	28.2	4,784	25.5
20 ~ 29	67	37.9	6,674	27.7	114	32.5	8,561	28.4	23	20.9	2,364	12.6
30 ~	11	6.2	857	3.5	79	22.5	2,975	9.8	16	14.6	908	4.8
Total	177	100.0	24,125	100.0	351	100.0	30,187	100.0	110	100.0	18,776	100.0

Source: Japan Paper Association

EXHIBIT 3-30

CONCENTRATION IN THE JAPANESE PAPER INDUSTRY

Changes in Shares of Top Companies in Production
of Main Pulp and Paper Grades 1960 - 1962

Pulp

Grades	No. of top Companies	1960	1970	1975	1979	1980	1981	1982
	1	11.1%	14.1	13.0	15.0	14.5	14.4	14.5
	3	27.8	36.1	36.0	36.4	34.1	34.8	34.9
Total Pulp	5	40.0	47.2	53.0	54.3	52.9	53.8	53.4
	10	60.8	65.9	70.1	72.1	70.9	71.9	72.1
	1	36.6	44.4	37.3	41.2	44.1	42.5	44.6
	3	71.0	91.2	100.0	100.0	100.0	100.0	100.0
Dissolving Pulp	5	94.5	100.0*					
	6	100.0						
	1	11.6	11.2	16.4	17.2	16.6	16.0	16.3
	3	31.2	28.0	35.2	43.2	41.8	40.6	40.3
BL Sulfate Pulp	5	49.0	43.8	52.6	59.5	59.0	59.3	58.6
	10	79.3	72.6	81.5	81.4	81.3	83.9	83.5
	1	12.3	17.5	14.7	15.6	16.1	16.0	16.9
	3	35.7	42.6	41.6	42.6	43.6	43.6	42.3
Unbl. Sulfate Pulp	5	52.5	62.0	58.5	60.1	61.9	61.3	59.9
	10	79.8	83.6	82.2	85.2	87.2	86.0	86.1

Top 4 companies

Cont. EXHIBIT 3-30

Paper								
	1	15.5	16.0	15.7	17.7	17.3	16.6	16.5
	3	36.5	39.0	37.4	38.3	37.3	37.4	36.8
Total Paper	5	47.3	47.6	48.9	51.1	50.8	51.4	50.9
	10	64.1	61.6	65.9	67.5	69.2	69.7	69.2
	1	29.6	28.6	29.3	29.1	28.2	28.2	28.5
	3	58.4	68.2	67.2	67.4	63.9	64.0	63.1
Newsprint	5	69.6	80.2	83.7	83.5	83.4	82.6	81.9
	10	85.9	96.7	98.1	98.2	98.2	98.0	97.8
	1	17.4	19.6	17.3	18.3	17.6	16.2	16.0
	3	36.2	44.2	40.2	43.6	42.3	41.8	41.7
Printing & Writing	5	51.1	60.1	57.9	58.6	57.8	57.8	57.4
	10	71.7	83.2	84.2	83.7	83.3	84.6	83.8
	1	17.5	30.2	25.0	24.2	23.6	22.4	22.5
	3	44.6	62.8	57.1	52.1	51.1	50.4	50.9
Unglazed Kraft	5	66.9	85.4	83.0	75.0	74.5	75.0	74.8
	10	98.0	100.0*	100.0**	100.0*	100.0*	100.0*	99.3

*Top 9 companies **Top 8 companies

Paperboard								
	1	8.4	9.8	8.6	7.1	6.8	6.5	6.7
	3	10.6	23.1	21.0	18.3	18.2	18.1	17.8
Total Paperboard	5	27.4	30.5	29.6	26.9	26.7	26.9	26.0
	10	42.0	45.2	46.1	43.8	42.3	41.1	40.8
	1	15.5	13.3	11.3	10.9	10.8	9.7	9.9
	3	32.2	32.6	29.9	28.5	26.2	24.2	24.2
Linerboard	5	44.3	46.5	44.8	41.6	39.7	36.5	36.0
	10	66.0	72.7	68.3	65.4	63.3	60.3	59.3
	1	14.4	12.4	12.6	15.1	16.1	14.9	12.8
	3	33.4	29.6	30.2	34.5	35.1	33.3	29.7
Corrugating Medium	5	47.5	44.0	43.2	50.8	52.5	50.2	45.0
	10	71.7	70.3	67.1	74.5	71.6	74.1	67.6
	1	29.1	18.7	14.7	13.2	12.9	14.0	14.0
	3	50.5	42.4	37.6	36.9	37.1	37.3	36.5
White Board	5	59.1	55.6	49.9	47.4	48.0	49.6	50.2
	10	72.8	71.2	67.2	65.7	65.1	65.5	66.4

Source: Japan Paper Association

EXHIBIT 3-31
Mitsui Group

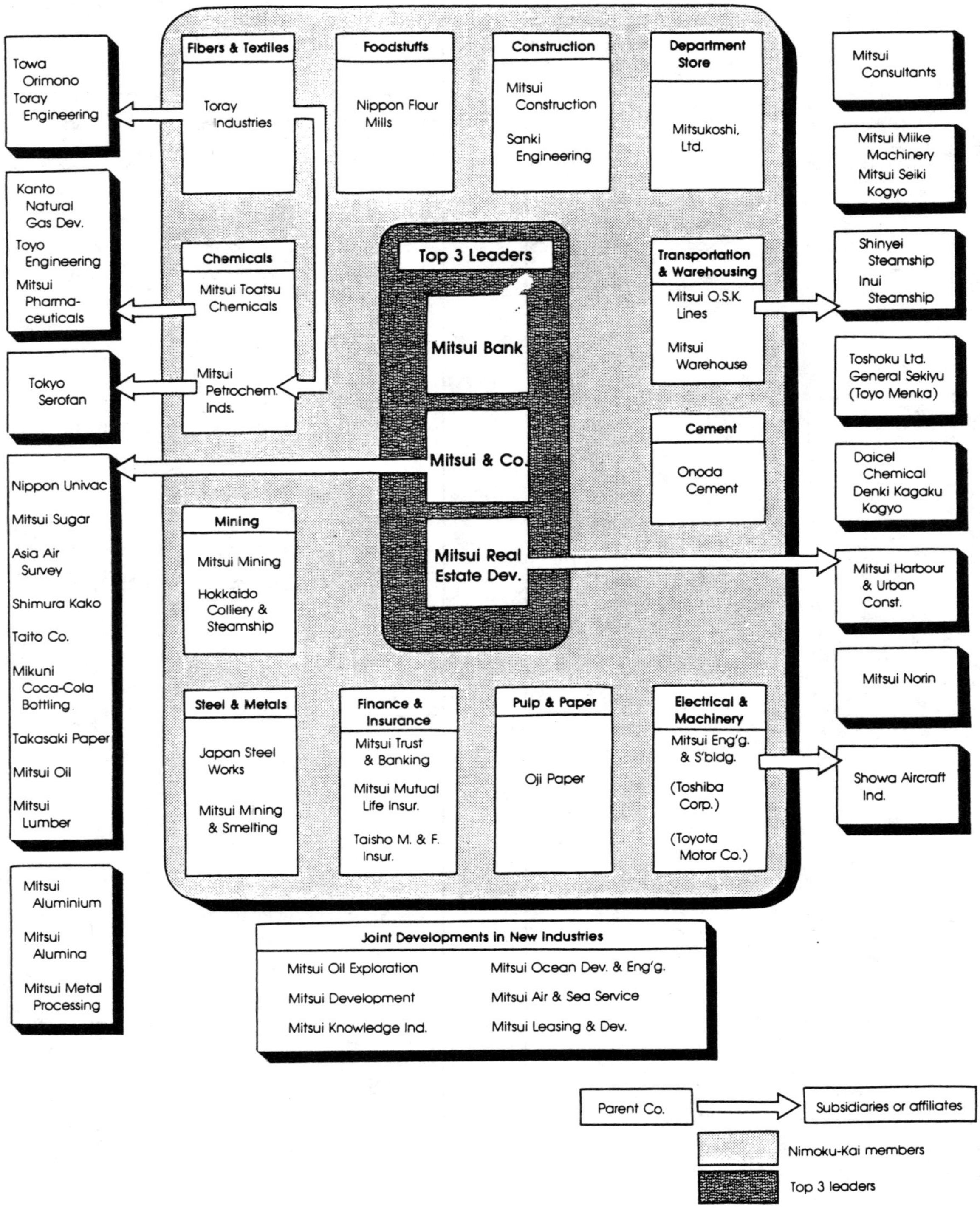

SOURCE: Daiwa Analysts Guide

EXHIBIT 3-32

Pulpwood Imports by Origins and Species 1960 - 1982

(In 1000 cubic meters)

Year	U.S.A.	Canada	U.S.S.R.	Southeast Asia and Other						TOTAL	Percent of total pulpwood supply (%)
				Australia	Malay-sia	New Zealand	Indo-nesia	Other	Total		
1960											
Total	–		192						1	193	2
of which softwood	–		173						1	174	3
of which hardwood	–		19						–	19	0
1965											
Total	260		179	–	26	–	9	–	35	474	3
of which softwood	260		157	–	–	–	0	–	0	417	5
of which hardwood	–		22	–	26	–	9	–	35	57	1
1970											
Total	4,282		276	–	572	158	25	172	927	5,485	20
of which softwood	4,244		134	–	–	153	–	1	154	4,532	39
of which hardwood	38		142	–	572	5	25	171	773	953	6
1975											
Total	7,784		696	2,033	702	394	88	171	3,388	11,868	41
of which softwood	7,196		422	–	–	363	–	–	363	7,981	55
of which hardwood	588		274	2,033	702	31	88	171	3,025	3,887	27
1979											
Total	8,407	836	802	3,315	281	503	163	547	4,809	14,854	45
of which softwood	7,783	836	598	–	–	389	–	1	390	9,607	57
of which hardwood	624	–	204	3,315	281	114	163	546	4,419	5,247	32
1980											
Total	8,103	1,249	589	3,684	340	499	140	835	5,498	15,439	45
of which softwood	7,307	1,249	431	–	–	408	–	38	446	9,433	54
of which hardwood	796	–	158	3,684	340	91	140	797	5,052	6,006	36
1981											
Total	6,130	906	588	3,208	262	350	148	588	4,556	12,180	42
of which softwood	5,383	870	464	107	1	285	–	16	409	7,126	50
of which hardwood	747	36	124	3,101	261	65	148	572	4,147	5,054	35
1982											
Total	5,347	751	495	3,390	226	432	112	755	4,915	11,508	40
of which softwood	4,842	685	363	84	–	351	–	25	460	6,350	45
of which hardwood	505	66	132	3,306	226	81	112	730	4,455	5,158	36

Source: *Japan Paper Association*
Note: *Canada is included in U.S.A. up to 1975.*

EXHIBIT 3-33

Japanese Fiber Imports
(,000 metric tons)

	PULPWOOD			PULP			WASTEPAPER		
	Imports	Consumption	I/C	Imports	Consumption	I/C	Imports	Consumption	I/C
1960	193	11,215	2%	147	3,679	4%	27	4,286	1%
1965	461	16,613	3	-	-		76	7,108	1
1970	5,485	27,901	20	917	9,718	9	110	12,260	1
1975	11,868	28,787	41	1,035	9,665	11	121	13,338	1
1979	14,854	32,897	45	1,868	11,861	16	131	17,810	1
1980	15,439	34,257	45	2,216	12,004	18	224	17,493	1
1981	12,180	28,711	42	1,698	10,310	16	81	16,998	1
1982	11,508	28,532	40	1,754	10,381	17	94	17,375	1

Source: Japan Paper Association, Pulp & Paper Statistics, 1983.

EXHIBIT 3-34

Japanese Pulpwood Imports as a Percent of Total Supply

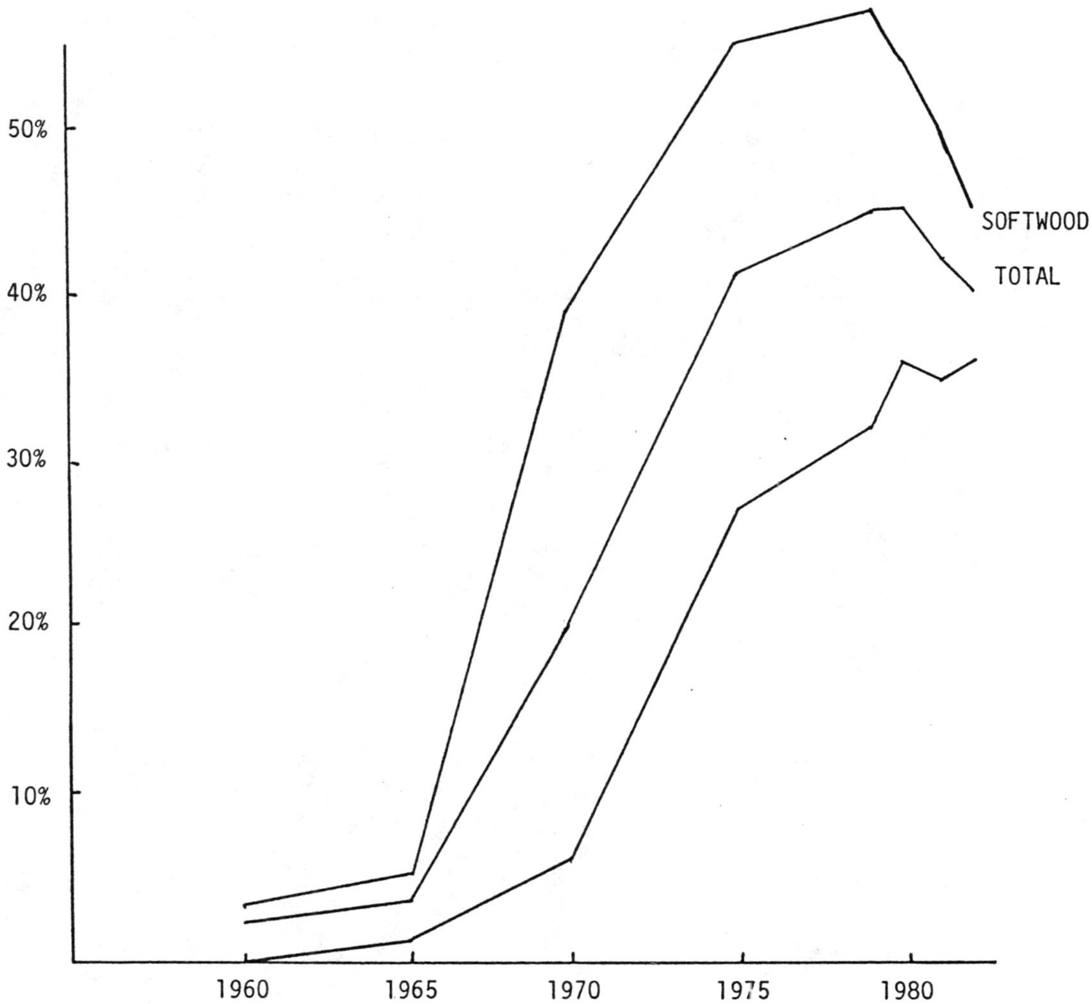

SOURCE: Japan Paper Association, Pulp and Paper Statistics, 1983.

EXHIBIT 3-35

Capital Costs of Pulp and Paper Mill, 1978

Pulp Making		
Pulping and recovery	$120 million	60%
Stock preparation	10 million	5
Paper Making		
Web forming	20 million	10
Wet processing	10 million	5
Drying	30 million	15
Finishing	10 million	5
		100%

Source: Strange, 1978.

EXHIBIT 3-36

Southern Softwood Pulp Production Costs
($/short ton)

	1976	%	1982	%	Compound annual 1976 % change 1976-82	1987e	%	Compound annual 1982 % change 1982-87
Freight CIF N. Europe*	42		56		4.9	77		
Variable costs:								
Wood	85	27	115	23	5.2	153	23	6.5
Labor	30	10	43	9	6.2	62	9	7.5
Energy	20	6	45	9	14.5	60	9	6.0
Chemicals	35	11	45	9	4.3	63	9	7.0
Other	30	10	55	11	10.6	74	11	6.0
Total variable costs	242	78	359	72	6.8	494	72	6.6
$ Overhead	13	4	23	5	10.0	32	5	6.8
Fixed costs:								
Interest	40	13	65	13	8.4	90	13	6.7
Depreciation	15*	5	55	11	24.0	75	11	6.4
Total costs	310	100	502	100	8.4	691	100	6.6
Consumer price index	170.5		289.1		9.2			

*Companies with long-term shipping contracts benefit from lower transportation rates.

Source: H. C. Bowen Smith, "US South pulp output costs analyzed," Pulp and Paper International, September 1983.

EXHIBIT 3-37

Cost of Building Extra Softwood Pulp Capacity

Company	Completion date	Capital cost ($ millions)	Capacity (tons/day)	Cost to make 1 ton/day ($1,000)
Great Lakes Forest	1966	42	520	80
Domtar	1967	75	650	115
Intercontinental	1968	70	570	120
British Columbia Forest	1972	65	520	125
Weyerhaeuser	1972	135	1,000	135
Procter & Gamble	1973	80	650	123
Alabama River Pulp	1978	285	1,000	285
Northwood Pulp & Timber	1982	260	830	313
Great Northern Nekoosa	1984	540	1,100	491
Champion International*	1986	540	750	720

*Hardwood

Source: H. C. Bowen Smith, "US South pulp output costs analyzed," <u>Pulp and Paper International</u>, September 1983.

EXHIBIT 3-38

Wholesale and Producers Price Indices for Selected Pulp and Paper Products, 1967 - 1982

	Bleached Sulfate Pulp	Waste Paper	Book Paper Uncoated offset	Kraft liner-board 42 lb.	News-print	Paperbags shipping sacks	Sanitary Paper Products	All SIC 26	Whole Sale Price Index
1967	100	100	100	100	100	100	100	100	100
1968	100	129	102	94	101	94	104	101	103
1969	100	140	106	98	104	98	109	104	106
1970	110	136	112	99	108	107	115	108	110
1971	112	112	116	100	112	100	120	110	114
1972	111	132	119	103	117	108	121	113	118
1973	128	223	121	111	122	119	125	122	126
1974	220	251	154	144	151	139	156	152	154
1975	281	90	158	171	184	156	195	170	172
1976	283	196	165	174	198	164	214	179	183
1977	277	184	174	171	216	172	237	186	194
1978	257	202	188	172	226	191	254	196	209
1979	308	278	214	196	250	220	280	219	236
1980	379	192	239	231	279	246	325	249	267
1981	396	137	266	252	308	NA	347	274	293
1982	375	95	247	247	316	263	354	289	299

Source: API, 1983 Statistics of Paper, Paperboard and Woodpulp, pp. 64-65

EXHIBIT 3-39

Variable Costs in SIC 26
($ millions and percent of value of shipments)

	SIC 26	SIC 2611 Pulp Mills	SIC 2621 Paper Mills	SIC 263 Paperboard Mills	SIC 2653 Corrugated Mills
Materials costs	30,148	1,212	7,222	4,050	4,661
(%)	(58)	(58)	(57)	(57)	(63)
All employees payroll	8,943	293	2,167	1,176	1,375
(%)	(17)	(14)	(17)	(17)	(19)
Production wages	6,323	223	1,632	888	867
(%)	(12)	(11)	(13)	(12)	(12)
Value of shipments	52,086	2,091	12,613	7,124	7,351
Value added	22,171	906	5,407	3,081	2,704
(%)	(43)	(43)	(43)	(43)	(37)

Source: 1977 US Census of Manufactures.

EXHIBIT 3-40

Average Hourly Earnings of Manufacturing
Production Workers in Selected States

	Georgia	Louisiana	Wisconsin	Michigan	Washington	Oregon
1950	1.08	1.25	1.47	1.72	1.73	1.79
1955	1.34	1.66	1.92	2.24	2.17	2.26
1960	1.66	2.12	2.37	2.75	2.63	2.55
1965	2.01	2.55	2.75	3.22	3.09	2.94
1970	2.67	3.28	3.61	4.15	4.06	3.82
1975	3.88	4.88	5.26	6.15	5.79	5.53

Source: BLS: Handbook of Labor Statistics

EXHIBIT 3-40

Average Hourly Earnings of Manufacturing Production Workers in Selected States

	Georgia	Louisiana	Wisconsin	Michigan	Washington	Oregon
1950	1.08	1.25	1.47	1.72	1.73	1.79
1955	1.34	1.66	1.92	2.24	2.17	2.26
1960	1.66	2.12	2.37	2.75	2.63	2.55
1965	2.01	2.55	2.75	3.22	2.09	2.94
1970	2.67	3.28	3.61	4.15	4.06	3.82
1975	3.88	4.88	5.26	6.15	5.79	5.53

Source: BLS: Handbook of Labor Statistics

EXHIBIT 3-41

Hourly Wages

	Paper and allied SIC 26	Paperboard SIC 263	Containers and boxes SIC 265	All manufacturing	Nondurables manufacturing
1965	$2.65	$2.93	$2.47	$2.61	$2.36
66	2.75	3.06	2.56	2.72	2.45
67	2.87	3.19	2.66	2.83	2.57
68	3.05	3.37	2.82	3.01	2.74
69	3.24	3.61	3.00	3.19	2.91
1970	3.44	3.86	3.18	3.36	3.08
71	3.67	4.14	3.40	3.57	3.26
72	3.95	4.47	3.64	3.81	3.47
73	4.20	4.78	3.88	4.08	3.68
74	4.53	5.14	4.16	4.41	3.99
1975	5.01	5.74	4.59	4.81	4.35
76	5.47	6.40	4.95	5.19	4.68
77	5.96	6.94	5.36	5.63	5.07
78	6.52	7.67	5.89	6.17	
79	7.13	8.38	6.37	6.70	
1980	7.84	9.28	6.94	7.27	
81	8.60	10.23	7.53	7.99	
82	9.32	11.11	8.17	8.50	

Indices: 1965 = 100

	SIC 26	SIC 263	SIC 265	All manufacturing	Nondurables manufacturing
1965	100	100	100	100	100
66	104	104	104	104	104
67	108	109	108	108	109
68	115	115	114	115	116
69	122	123	121	122	123
1970	130	132	129	129	131
71	138	141	138	137	138
72	149	153	147	146	147
73	158	163	157	156	156
74	171	175	168	169	169
1975	189	196	186	184	184
76	206	218	200	199	198
77	225	237	217	216	215
78	246	262	238	236	
79	269	286	258	257	
1980	296	317	281	279	
81	325	349	305	306	
82	352	379	331	326	

Source: API, Statistics of Paper, and 1977 Census of Manufactures.

EXHIBIT 3-42

	SIC 26 Avg. Hourly Wage	&	Index	Production output/hr (All employees)
1964	$2.56		42.3	62.3
1965	2.65		44.5	65.4
1966	2.75		46.1	68.7
1697	2.87		48.2	67.6
1968	3.05		51.2	71.9
1969	3.24		54.4	74.8
1970	3.44		57.7	77.5
1971	3.67		61.6	81.0
1972	3.95		66.3	87.9
1973	4.20		70.5	91.5
1974	4.53		76.0	91.5
1975	5.01		84.1	86.7
1976	5.47		91.8	95.0
1977	5.96		100.0	100.0
1978	6.52		109.4	103.2
1979	7.13		119.6	105.4
1980	7.84		131.5	105.4
1981	8.60		144.3	107.5
1982	9.32		156.4	108.3

Source: API, 1983 Statistics of Paper, Paperboard and Wood Pulp p. 54, 56

EXHIBIT 3-43

Workers Employed in Paper Manufacture and All Manufacture
by Country (000 workers)

	United States			Canada			FR Germany			Finland			Sweden		
	P&P	All Mfg	%	P&P	All Mfg	%	P&P	All Mfg	%	P&P	All Mfg	%	P&P	All Mfg	%
1973	705	20,142	3.50	118	1,594	7.40	556	9,211	6.04	50	509	9.82	47	650	7.23
1977	692	19,682	3.52	125	1,545	8.09	458	8,340	5.49	51	503	10.14	48	634	7.57
1980	693	20,285	3.42	125	1,614	7.74	483	8,433	5.73	51	533	9.57	47	602	7.81
1981	689	20,170	3.42	127	1,624	7.82	481	8,195	5.87	50	530	9.94	46	579	7.94
1982	662	18,853	3.51	119	1,472	8.08	467	7,893	5.92	na	na	na	na	na	na

Source: ILO, Yearbook of Labor Statistics.

EXHIBIT 4-43

Workers Employed in Pulp and Paper
As Percent All Manufacturing

	U.S.	Canada	FR Germany	Finland	Sweden
1973	3.50%	7.40%	6.04%	9.82%	7.23%
1977	3.52	8.09	5.49	10.14	7.57
1980	3.42	7.74	5.73	9.57	7.81
1981	3.42	7.82	5.87	9.94	7.94
1982	3.51	8.08	5.92	na	na

Source: ILO, *Yearbook of Labour Statistics*.

EXHIBIT 3-44

PULP, PAPER & PAPERBOARD INDUSTRY

Changing Fuel Mix, 1972-1982

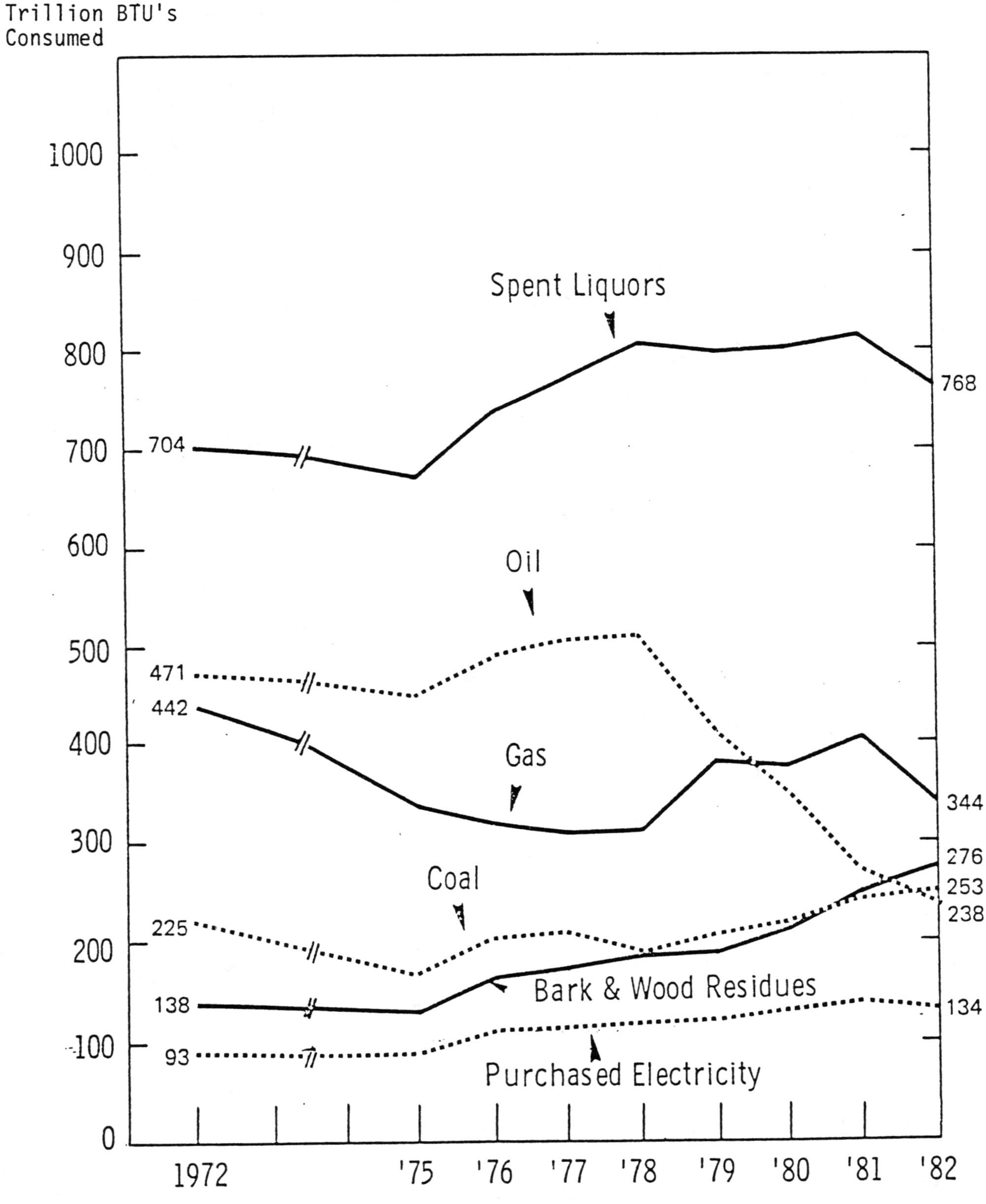

Source: API, March 1983

EXHIBIT 3-45

U.S. Pulp and Paper Industry
Energy Uses by Fuel Type (1982)

Spent pulping liquors	37.4%
Natural gas (P)	15.9
Bark and log fuel	13.4
Coal (P)	12.3
Fuel oils (P)	11.6
Purchased electricity (P)	6.5
Other (P)	2.9
	100.0%

(P) = Purchased Fuel
Source: Standard and Poors

EXHIBIT 3-46

Energy Use in Pulp and Paper Making

Process	Energy Btu's/ton
Pulp	
Pulping & recovery	15 million
Stock preparation	4 million
Papermaking	
Web forming	3 million
Wet pressing	3 million
Drying	6 million
Finishing	0.3 million
Total	31.3 million Btu's/ton

Source: Strange, 1978

EXHIBIT 3-47
Energy And Papermaking: Purchased And Self Generated 1972-1981

FIGURE 1A: Fossil fuel and purchased energy consumption (adjusted)/ton
FIGURE 1B: Growth in paper industry's self-sufficiency as percent of total

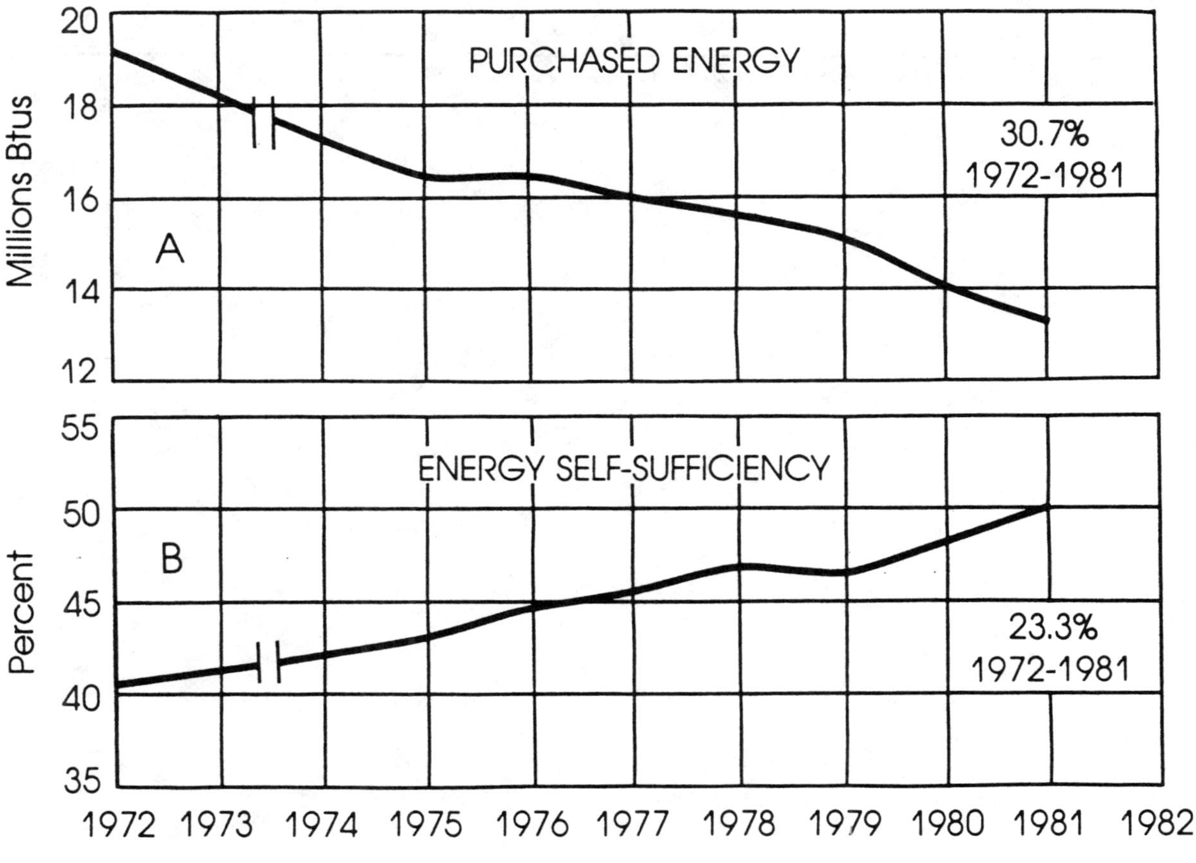

SOURCE: John C. W. Evans, "Industry makes significant gains in energy use during past 10 years." Pulp and Paper, November 1982, p.96.

EXHIBIT 3-48

R & D Expenditures

	Paper	Chemicals	All Industry
R & D Expenditures ($ billion)			
1982	.272	3.055	36.200
1978-82 %	+8.0	+15.7	+16.0
As % of sales			
1982	1.0	2.9	2.4
1980	0.9	2.4	2.0

Source: Business Week, 3/21/84.

EXHIBIT 3-49

Research and Development
U.S. Pulp and Paper Industry Expenditures
1966 - 1983e

	$ million	% Sales
1966	117	0.7
1967	128	0.8
1968	144	0.8
1969	188	0.9
1970	178	0.9
1971	187	0.8
1972	189	0.8
1973	194	0.7
1974	237	0.8
1975	249	0.8
1976	313	0.8
1977	336	0.8
1978	391	0.8
1979	454	0.8
1980	508	0.9
1981	557	0.9
1982	586	1.0
1983 est	597	n.a.

Source: API, Statistics of Paper, 1983

EXHIBIT 3-50

U.S. Pulp and Paper Industry
Expenditures for Environmental Protection

	Expenditures ($ million)	As % of Sales	As % of Total Capital Expenditures
1966	409	2.4	
1967	76	0.4	
1968	93	0.5	
1969	128	0.6	
1970	187	0.9	
1971	203	0.9	
1972	339	1.5	38
1973	351	1.3	34
1974	523	1.7	31
1975	645	2.0	30
1976	564	1.4	29
1977	443	1.1	19
1978	363	0.7	16
1979	359	0.6	12
1980	368	0.6	9
1981	260	0.4	6
1982	318	0.5	9
1983 est	323	n.a.	n.a.

Source: API, Statistics of Paper, 1983

EXHIBIT 3-51

Wood Pulp Capacities in Key Regions: 1970
Through Forecast 1991
1,000 metric tons

(figures in parenthesis are % change from previous figure)

	U.S.	Canada	Japan	EC[1]	Scans[2]	Oceania[3]
1970 (% change)	37,640 (+12)	18,842 (+15)	9,122 (+30)	5,714 (+10)	19,290 (+5)	1,141 (+81)
1975	42,219 (+11)	21,656 (-)	11,890 (+6)	6,265 (-1)	20,290 (-1)	2,064 (+8)
1980	47,104 (+11)	21,524 (+9)	12,573 (-3)	6,203 (+1)	20,085 (+2)	2,226 (+19)
1985 est[4]	52,501	23,362	12,231	6,265	20,565	2,660
1987 est	47,679	22,785	13,490	6,792	22,370	3,804
% Δfrom 1975	(+13)	(+10)	(+13)	(+8)	(+10)	(+84)
1991 est[5]	52,911	23,559	13,060	6,723	21,643	3,316
% Δfrom 1980	(+120	(+9)	(+4)	(+8)	(+8)	(+49)

Source: [4]OECD <u>Pulp and Paper Industry</u>;

[5]FAO, <u>Projected Pulp and Paper Mills in the World</u>.

[3]Australia, New Zealand, Papua New Guinea

[2]Norway, Sweden and Finland

EXHIBIT 3-52

Paper and Paperboard Capacities in Key Regions: 1970 Through Forecast 1991

1,000 metric tons (figures in parenthesis are % change from previous figure)

	U.S.	Canada	Japan	EC	Scans	Oceania
1970	50,839	12,765	13,995	23,679	10,830	11,619
% change	(+12)	(+12)	(+41)	(+13)	(+28)	(+20)
1975	57,172	14,218	19,800	26,714	13,915	1,936
	(+9)	(+3)	(+9)	(+5)	(+9)	(+22)
1980	62,491	14,682	21,602	27,996	15,160	2,367
	(+10)	(+13)	(+7)	(+6)	(+14)	(+17)
1985 est[1]	68,431	16,654	23,705	29,706	17,300	2,767
1987 est[1]	62,429	14,166	21,674	27,916	16,908	2,294
% from 1975	(+9)	(0)	(+9)	(+4)	(+22)	(+18)
1991 est[2]	68,726	16,537	23,856	29,640	18,038	3,597
% from 1980	(+10)	(+13)	(+10)	(+6)	(+19)	(+52)

Source: [1]OECD, The Pulp and Paper Industry, [2]FAO, Projected Pulp and Paper Mills in the World.

CHAPTER 4

INTERNATIONAL TRENDS IN GOVERNMENT POLICIES AND REGULATIONS

Apart from national differences in demand and supply conditions, industry structure, and factor endowments, different government policies and regulations can provide global or, on a smaller scale, regional or national competitive advantages or disadvantages to different national industries. This chapter traces the trends in such policies and regulations in as far as they affected competition in the global pulp and paper industry in the 1970's and early 1980's. Certain regulatory trends encompassed environmental issues such as pollution regulations, forest-use regulations, and recycling and energy policies. Other forms of government intervention were more internationally oriented, such as subsidies to the industry, industry restructuring efforts, and international trade issues such as tariffs and quotas, dumping and price fixing, and the effects of certain key and controversial currency devaluations. After a brief description of the methodology used for this section, the chapter first summarizes the US industry's concerns with government policy, then examines world trends in government policies and regulations impacting directly or indirectly, positively or negatively, on the leading nations' pulp and paper industries.

METHODOLOGY

The main focus of the research methodology was a content analysis of industry sources from 1974 through early 1984. Beginning with the US, available literature and journals were examined as well as the annual reports of 19 US companies from 1970 to 1982. The second stage of the content analysis was concerned with the major European trade journal, Pulp and Paper International, supplemented by national journals where available. This analysis involved identifying issues and trends over the past ten years. Furthermore, because of the significant impact of the US industry globally, this source--internationally oriented as it was--proved the best source as far as US-international issues were concerned. Finally, a number of US industry and government representatives were interviewed, the results of which were added to insights from other members of the CIPS team who were involved in the research project in North America, Western Europe, and Japan.

GOVERNMENT CONCERNS OF THE UNITED STATES PULP AND PAPER INDUSTRY

Government concerns of US corporations tended to be parochial. Exhibit 4-1 shows the number of times government-related issues were mentioned in the study's survey of US corporations' annual reports. These issues break down roughly between environmental issues and, as required by law, the reporting of antitrust or price-fixing investigations concerning the corporations. Only three times in 13 years do any of the 19 companies surveyed mention an internationally oriented business-government issue. This relative disinterest in international issues is likely explained by the relative strength globally of the US pulp and paper industry in contrast to the US textile industry, for example, which appears fascinated with what the US government should do for them and what other governments should not be doing for their textile industries.[1] It is also in sharp contrast to the concerns of non-US pulp and paper industries which, in addition to concerns similar to the US industry concerning environmental policies and price fixing, were also far more concerned with international competitiveness issues and how their countries' government policies were impacting this situation.

WORLD TRENDS IN GOVERNMENT CONCERNS WITH INDUSTRY-RELATED ISSUES

On a global perspective, certain trends appear in governments' concerns with pulp and paper industry-related issues. In the early 1970's, environmental issues became the major concern worldwide. Environmental issues included a new emphasis on pollution control as well as concerns of both governments and industries as to diminishing natural resources, particularly forests and energy sources. The improvement of pollution control within the pulp and paper industry coupled with the world recession following the oil-price increases in the mid-1970's caused environmental issues to drop in relative concern. In their place arose issues of government subsidies and industry restructuring plans for "threatened" industries and "unfair" and protectionist trade practices (see Exhibit 4-2). Each of these issues--environmental, government subsidies, and international trade issues--are examined in more detail below.

ENVIRONMENTAL ISSUES

Environmental issues in the early 1970's greatly affected the global pulp and paper industry. Increased government regulations aimed at pollution control significantly increased capital costs for the industry. Increased awareness that natural resources were being depleted stimulated interest in policies concerning forests and the recycling of wastepaper. Finally, increased energy costs and subsequent national policies concerning energy affected the pulp and paper industry which is a heavy consumer of energy. While industries in each country tended to see environmental policies as a problem more or less unique to themselves, these problems were essentially worldwide although the timing and magnitude of the policies and their impact varied from country to country.

Pollution Regulations

In the first half of the 1970's, a proliferation of pollution control laws began to affect the pulp and paper industry in the US and other major producing countries.

In the US, the major pieces of legislation that affect producers were the Clean Air Act of 1970, amended in 1977, and the 1972 Federal Water Pollution Control Act with its revision, the Clean Water Act of 1977. In addition, the use of chemicals in forestry came under greater scrutiny and control.[2] Such legislation was spurred on by consumer advocates such as Ralph Nader's Study Group whose report in 1974 lambasted pulp and paper producers and admonished government agencies for not having done enough to protect the environment from the polluting effects of this industry.[3] Early reaction to this new legislation can be characterized by a comment made by J. P. Tonelli, president of United Paperworkers International Union. Speaking at the 96th Annual Paper Week sponsored by the American Paper Institute (API) in 1973, he remarked, "The public has a right to be concerned with the environment. Our industry, too, must be concerned with its shareholders and worry about its rate of return on investment."[4] This attitude of desiring to be a good citizen combined with a great concern with the cost of being one is seen again in our study's survey of US corporate annual reports. Compliance with environmental protection legislation was the second most often mentioned concern with government regulations, and the references often took a similar tone as Mr. Tonelli's remark, as depicted in Exhibit 4-1.

Indeed, costs for pollution control loomed large for the US industry. A spokesperson for the 1973 API meeting told a press conference that the US paper industry alone planned to spend close to $1 billion on controlling air and water pollution in 1973-1974, and that this represented at least 50 percent of the industry's total planned capital investment. These new figures compared with a combined pulp and paper industry investment of about $1,110 million on air and water control up through 1971.[5] While the initial costs of dealing with new legislation were high, they did decrease in subsequent years. A Pulp and Paper study of seven-year capital spending trends revealed that the US pulp and paper industry's environmental spending was 25 percent of total capital spending in 1977 as the industry strove to meet the new 1977 standards. The following years, however, saw a decline to around 10 or 11 percent of total allotted to environmental control with a slight increase in 1982 to 13 percent as a result of new 1983 air-quality standards.[6]

Despite the decrease in environmental spending after the initial gargantuan investment, the US pulp and paper industry undeniably took on a great added expense complying with the new antipollution legislation. They were not alone, however. Despite inherent problems in comparing the effects of different national environmental policies and their enforcement, one can determine that no other major pulp and paper producing country was immune to the new antipollution regulation sweeping the globe by the mid-1970's. Even developing countries found themselves contending with these issues, albeit to a lesser extent than the developed nations.

Looking first at Canada, one sees that spending on pollution control in the 1970's and early 1980's was substantial there as well. Environmental spending averaged 13 percent or 14 percent in 1977 and thereafter except for 16 percent in 1978 and 12 percent in 1980.[7]

In Europe, the EC, and subsequently pulp and paper producers within the participating countries, began to deal with antipollution legislation concurrently with the US. By 1973, the Commission of the Common Market in Brussels was working on an all-encompassing environmental protection law and was paying particular note to the pulp and paper industry. However, the problems of how to unify the various emerging regulations of different member states as well as how to balance various national incentives aimed at assisting industry to comply to these regulations remained a formidable task.[8] A lack of uniformity, however, did not stop the rising costs associated with new national legislation within the EC and the other European states. In an interview with Pulp and Paper International in January 1976, Claus Koren, chairman of the environmental committee of Norway's Pulp and Paper Association, responded to US complaints concerning antipollution laws, stating that the pulp and paper industry in his country was bearing just as great a burden in proportion to its production. He further stated that pollution restrictions within Norway would increase the amount of capital needed for a new integrated pulp and paper mill by around 10 percent and annual operating costs by about 5 percent--an estimate slightly lower than actual expenditures in the US and Canada.

Japan also experienced the phenomenon of dealing with antipollution legislation concurrently with the US. A study by Pulp and Paper International published in February 1981 shows the antipollution component of capital investment in Japan's pulp and paper industry was less than 6 percent in 1970, but rose to between 18 and 25 percent annually between the years 1972 and 1976. By 1978, however, it had dropped to about 4 percent, where it stayed through 1980.

In developing countries, such as Brazil, pollution was also a problem and a consideration for the pulp and paper industry. Although the Brazilian delegation to the United Nations Environmental Conference in Stockholm in 1972 declared that their country would not accept environment legislation that interfered with industrialization, some attempts to control pollution were made. From the early 1970's, new mills in Brazil were required to have basic chemical treatment recovery equipment, and mills near population centers were subject to stricter pollution controls.

Therefore, the movement toward environmental protection affected all the major trading nations in pulp and paper. Furthermore, there does not appear to be a significant lag between the time this movement hit one country or another excepting the developing world, where industrial development will likely continue to be weighed higher than pollution control in the foreseeable future. Major investments by the world's pulp and paper industries to comply with new environment legislation was essentially absorbed in the mid-1970's and has fallen as a percentage of total annual capital investment since then. However, evidence does indicate that not all national industries were required to invest similar amounts toward pollution control. As a result of different legislation, enforcement of that legislation, and/or different technology, such investments in Japan for example appear to fall distinctly below those of North America.

Forest Regulations

National regulation of forest lands and their uses can have a direct impact on a nation's pulp and paper industry. Concurrent with the world attention concerning the effects of pollution on the environment which arose in the 1970's came a concern in many nations relating to their diminishing pulp supplies. This concern manifested itself in issues ranging from the public versus private ownership of forests to reconsideration of tax laws affecting the use of forest lands.

In the US comprehensive land-use planning measures were proposed during several sessions of Congress over the past 15 years. Such legislation was aimed primarily at the owning, growing, and harvesting of trees and at establishing zoning of forest lands. After considerable controversy, the proposed comprehensive measures were tabled at the national level in the US, although several states adopted some form of land-use regulation.[9]

One piece of environmental legislation that was passed, however, was the Endangered Species Act in 1973. While this law did not immediately affect the US pulp and paper industry in its entirety, the American Pulpwood Association opposed the legislation on the ground that, "There is insufficient consideration under the Act of national goals and objectives other than conservation of listed species and their critical habitats."[10] The industry also opposed proposals in 1975 to list the Red Hills Salamander as an endangered species, claiming that doing so would adversely affect timber harvesting of 185 million feet of lumber and 800,000 cords of pulpwood annually. Opposition also arose to a proposal to list 13 million acres in the Western US as critical habitat for the grizzly bear. Similarly, the US industry opposed the Natural Diversity Preservation Act proposed in 1977 which aimed at creating an office within the Department of the Interior which would identify and protect "examples of all elements which collectively comprise the nation's natural diversity."[11]

Probably the most heated controversy in the US continuing into the 1980's surrounds the issue of public forest land withdrawal. Since the Wilderness Act of 1964, the US Congress continued to increase designated wilderness areas. Hopes for a comprehensive bill relating to the designation of all national forest lands was abandoned in 1980, giving way to many individual bills affecting a single state or national forest. This issue, the controversy surrounding it, and opposition originating in industries continues, including but not limited to the pulp and paper industry. The American Pulpwood Association, voicing the concerns of its constituents, declared, "The rapid rate of withdrawals from commercial forest land comes during a 30-year period in which the demand for forest products is expected to double. Without greatly increased productivity from the remaining commercial timberlands, critically high wood costs can be expected by the year 2000."[12]

Fear of diminishing timber resources had a more immediate effect in other countries during the 1970's. In the early 1970's Sweden was faced with an annual cut rate of timber which was thought to be exceeding the annual growth rate. This was particularly alarming for a nation as highly dependent on its pulp and paper industry as is Sweden. At the end of 1972 the Swedish government presented a comprehensive bill on physical planning setting forth guidelines on the balance of forestry and recreational activities and the location of pulp and paper mills. In cooperation with the government, the Joint Committee of the Swedish Forest Industries had presented their foreseeable requirement prior to the passing of the bill. The industry was caught by surprise, however, when the government temporarily halted any mill expansions in October 1973 when the Swedish pulp and paper industry was about to launch into an active expansion period.[13]

In Canada, government reassessment of its timber resources also took place. In late 1975, the Quebec government began buying back major public-forest concessions from private parties. Their avowed purpose on doing this was not to increase revenue but to better manage the province's wood resources. At the same time the Pearse Commission in British Columbia was studying that province's forest resources, causing pulp mill operators there to be concerned over reallocations of timberland.[14] The British Columbia Forest Ministry did in fact act, freeing chips from an old allocation system as of April 1980 and forcing pulp mills to compete against each other on the open market at a consequently sharp increase in price.[15]

Developing countries were also becoming more desirous of protecting their forest resources. Countries such as Chile and Brazil began to contend with reforestation by undertaking new programs emphasizing tax incentives. In East Asia, Japan, itself facing an inability to substantially increase its indigenous chips supply, began to face growing opposition from neighboring nations toward exporting chips to Japan; e.g., the Philippines forwarded a new policy forbidding the export of logs after 1976. The trend in many developing countries during this time was toward developing national pulp and paper industries. This trend was controversial. Some believed that certain large pulp-producing projects in developing countries such as Brazil, Morocco, and several West African states were too oriented toward marketing toward the developed world and could face fatal competition from North America.

While the access to and use of public forest lands present a number of policy issues, many forest lands are privately owned, and a major policy issue affecting these lands is that of taxation and forestry incentives.

In the US, 78 percent of timberland is privately owned, and only 20 percent of that is owned by the forest industry. To alleviate some of the potential problems encountered by these investors who face a long period between planting and harvesting, the US developed a program of relatively favorable tax legislation. Since 1921, growers have only been subjected to federal capital gains tax on their earnings from timber sales instead of being subjected to the higher personal income tax rates on these sales. In addition, many individual states have passed legislation aimed at reducing or eliminating state taxes on growers until timber is harvested and income is realized. Also, government assistance has been available, providing free or partially subsidized forest management services and fire protection, some examples being the Smith-Lever Act of 1914, the Clark-McNary Act of 1924, the Norris-Doxey Farm Forestry Cooperative Act of 1950, and the relatively recent Reforestation Tax Incentive of 1980. The US Forest Service is estimated to have spend millions of dollars over the past 25 years to encourage better forest management. The consensus is, however, that few privately owned nonindustrial forests in the US produce close to their potential. Some experts argue that this is not solely due to abundance and relatively low price of US timber, but is also due to the "favorable" tax policy for "hoarding."[16]

A different tax policy prevails in Finland where forest ownership breaks down similar to that in the US but where there is more immediate concern about the prospects of an adequate national timber supply. As in the US, government aid is available in Finland for initial planting, and while taxes are assessed annually, they may be deferred until the first commercial thinning. It is the nature of the annual taxes which differs significantly from those in the US. Finnish taxes are imposed on a presumed income per hectare which is based on (1) which one of the five capability classes the taxes property falls under and (2) a three-year average of prevailing local stumpage values. Therefore, there is an incentive to capture the most growth potential from forest lands since taxes are based on average yields. In the US, on the other hand, owners receive the same tax incentives for allowing natural regeneration to occur as for managing a capital-intensive forest.[17]

The US and Finland represent two extremes in tax policy. In countries like Finland where timber supply is becoming a more critical issue, one can expect to see more policies tending to forced well-managed growth. In an editorial in 1980, Pulp and Paper International urged other West European governments to move toward the Finnish style of tax system which "rewards the active manager and penalizes the hoarder."[18,19]

Policies Affecting Recycling of Wastepaper

Beginning with the ecological movement in the early 1970's, there was renewed interest in recycling wastepaper. In the US, the issue of reducing solid waste appears to have dominated over the issue of diminishing pulp supplies. In 1973, the Chief Executive Officer of Berstrom Paper Company in Wisconsin proposed a tax credit to encourage recycling, citing society's and subsequently government's desire to recycle in order to address a problem of pollution. The proposed tax credit of 18 percent on total annual dollar purchases of wastepaper was designed to place a company which was recycling at a parity with companies receiving their special capital-gains tax treatment for the trees they harvested.[20]

European governments, particularly those in France, Holland, Sweden, and the UK, showed greater concern with the issue of securing pulp supplies and began various measures to increase recycling in the mid-1970's. In May 1976, the French government launched a scheme of subsidies aimed at encouraging recycling in hopes of cutting pulp imports. While most of the aid was directed at investment in treatment plants, some also went toward wastepaper stock-building operations. Holland concurrently introduced plans to encourage wastepaper collection and storage.

The UK, on the other hand, provided grants totaling 23 million in 1977 to increase capital investment in recycling but balked at supporting the collection and storage side of recycling. Local authorities were bound by law to collect household waste but not required to separate wastepaper from other household rubbish. It was estimated that only one in three municipalities operated its collections in such a way that wastepaper could be recovered. In addition, private investors in wastepaper collection faced the discouraging hazard of relatively steep fluctuations in price tied to fluctuating pulp prices in a cyclical industry. Without a stock support system of a reliable supply, the UK found itself faced with sharp jumps in wastepaper imports at times of increased demand. While this was a short-term answer, it did nothing to forward the purposes of recycling, that is, cutting down both on solid waste and dependence on foreign supplies.

Sweden attacked the issue from the opposite point of view from Britain. Instead of concentrating on direct aid for capital investment in machinery and recycling plants, the Swedish government aimed at creating favorable conditions for such investment by guaranteeing increased and stable supplies of wastepaper. A new law which went into effect the first of July 1975 made municipal authorities throughout Sweden responsible for the separate collections from households of paper and other waste.[21] The system appears to have been successful. Considerable investment in recycling plant and equipment did indeed take place. In 1980, recovery of wastepaper from households was running just about 40 percent with hopes of raising it to 50 percent. Recycled paper consumption had doubled since

1974 and represented 11 percent of total consumption by the paper industry.

In Japan, similar government concerns about ecology and shared government and industry concerns about increasing import dependence on pulp also resulted in substantial efforts and success in the increased use of recycled wastepaper. In fact, of all major producing countries, Japan's use of recycled wastepaper as a percentage of fiber input (40 to 50 percent) exceeds all other countries. (The comparable percentage for the US is less than 20 percent.)

Thus the greatest concern on the part of governments toward actively becoming involved in recycling appears to have been inspired by worries concerning pulp supply, and the most advantageous way of addressing this concern appears to have been systems in which the issue of reliable wastepaper supply was not ignored.

Energy Policies

Another environmental issue affecting the global pulp and paper industry in the 1970's and early 1980's was the issue of energy prices. The significant jump in prices which occurred was particularly important to such an energy-intensive industry as pulp and paper. In Sweden, for example, the forest-related industries represented 40 percent of the country's entire industrial energy consumption.[22,23]

Despite the increase in prices, relatively cheaper energy remained available to the industry in North America due at least partially to government policies which in the eyes of Europeans favored cheap energy sales to industry. By the early 1980's, North America did indeed enjoy a cost advantage compared to other countries. An EC commission declared that Canadian papermakers paid half as much as their EC counterparts for oil and 59 percent less for electricity. Similarly, a typical integrated mill in Japan faced energy costs about three times higher than a comparable mill in North America.[24]

Within Europe itself controversy arose concerning energy policies and consequent prices. The UK, who even more than North America enjoyed independent and rich energy sources in the early 1980's, did not see these sources translated into lower energy costs for its pulp and paper industry. The British Paper Board Industry Federation claimed that heavy fuel oil and electricity prices were 20 percent higher than most of their EC competitor countries. This was apparently due to several factors. First, Germany, France, and Sweden made large government loans or grants available for efficiency measures in industry. Second, the British excise duty on heavy fuel oil was more than double that of other EC countries. Third, heavy users of electricity received greater discounts in Germany, Holland, Belgium, and Italy compared to England and Wales. Thus, a number of various governmental environmental policies such as these could and did influence the important energy cost factor within the global pulp and paper industry.

GOVERNMENT INTERFERENCE IN THE MARKET

While environmental concerns dominated business-government issues within the international pulp and paper industry in the early 1970's, a different category of concerns dominated the latter half of the decade. The oil price shock of 1973-74 and the consequent world recession of 1975 and 1976 profoundly affected an industry tied so directly to business cycles. As many companies worldwide but particularly in Europe faced bankruptcy, it was not surprising that a major issue of the second half of the 1970's was government subsidies to their pulp and paper industries.

In Europe, the question of what was the proper extent of governmental interference became a particularly controversial subject. In Sweden, a unique tax law helped alleviate some of the initial shock. The Swedish pulp and paper industry was allowed to set aside tax-free part of their profits during good years if they were subsequently reinvested in the business during lean years. As the effects of the recession continued, however, the Swedish government had to inject more direct and enormous financial aid in order to save companies--much to the consternation of their EC competition.[25]

The controversies which raged within the EC were even more heated and proved a hard test for international agreements among the member countries. Behind the controversy was a harsh reality. The need to consolidate and modernize in order to stay competitive resulted in a loss of 34,000 jobs within the EC industry between 1975 and 1980.[26] The threat of company closures proved too great a worry for national governments. Despite untested and to a degree ambivalent agreements not to interfere in national industries, numerous EC member governments interfered.[27]

An early form of interference was an investment subsidy from the British government. In April 1976, the government proposed to make 25 million available to paper producers for capital investment over the next five years, if such an action proved not to be in conflict with the EC rules forbidding aid which gives ailing enterprises an "unfair" competitive edge. This represented the first government grant ever to the British paper industry. By the next summer, 30 UK companies had applied for the aid, and the EC had not moved against it.[28]

In Belgium, Holland, and France aid was directed at saving specific companies. In the autumn of 1980, the Belgium government agreed to pay off $50 million in debts of Intermills, the country's leading producer of special printings and copy papers, and to guarantee new debt of over $30 million. Regional authorities injected an even higher amount as a shareholding. Similarly, the Dutch government invested $1.5 million in the Leeuwarden Papierwarenfabriek boxplant in northern Holland. The French government discussed the possibility of state interference to restructure their industry in 1977 and 1978, but settled on a massive aid plan instead. (The Japanese government was concurrently considering the same thing.) The plan consisted of partially subsidized loans of $73 million and $32 million to two larger companies and loans totaling $65 million to be split among three smaller countries.

These government actions, however, attracted controversy and the attention of the EC Commission's competition department, "DG4." In late 1982, DG4 called on the Belgium, Dutch, and French governments to explain and justify the aid they had been giving to their paper industry and to individual companies in particular. In the cases of Belgium and Holland, the respective governments were asked to analyze the market effects of their actions. Competing producers would in turn comment on the governments' analyses, with the EC Commission arbitrating if necessary.[29]

These actions by the DG4 aroused varied, sometimes strong sentiments. National industries with the most to gain from government aid on the whole denounced the actions, professing that industries in non-EC competitor countries such as Sweden, Austria, and Canada were receiving more aid.[30] On the other hand, the Federal Republic of Germany, who was faring the best of the EC pulp and paper industries, decried government interference and supported the actions of the EC Commission.

The success the EC Commission will have in seeing its decisions actually transformed to actions within member nations remains to a degree untested. The Commission's decision against Holland and Leeuwarden Papierwarenfabriek in mid-1982 resulted in the Dutch government's announcement in 1983 of its determination to test the decision in the European Court of Justice. The government claimed that only a small percentage of the plant's output of converted board was exported and as such did not affect EC trade. It also claimed that the aid to the company contributed to the development of a depressed region, an allowable act under EC rules. Nonetheless, the DG4 was becoming if anything more active, despite alleged government attempts to find loopholes in laws and sidestep regulations. In late 1982, the competition department announced its opposition to Italian plans for aid to its industry in general.

In Canada, between 1973 and 1976, almost all major mill constructions received substantial assistance from federal or provincial governments. Furthermore, during this time, provincial governments in Canada took over several pulp and lumber companies. Manitoba invested more than $90 million in the former Churchill Forest Industries, Ltd., forming Manitoba Forest Resources, Ltd. When Crown Zellerback planned to close a mill in British Columbia in 1973, the province took it over rather than see it close resulting in the loss of hundreds of jobs. Several years later, however, planners estimated that the mill needed an injection of $50 million to make it competitive, and the government balked, permitting instead $2 million to perform immediately necessary improvements. The controversy of subsidies versus the free market in Europe was echoed by British Columbia's premier Bill Bennett stating the "government can't be in business where it is controlling the game and also allowing the private sector to operate."[31] Still, the specter of bankruptcies and mill closings in a major industry was as threatening in Canada as it was in some European countries.

In Japan, domestic and international developments combined to throw the industry into such a severe recession that the industry sought government intervention. The 67 percent increase in the price of imported wood chips in late 1979 (euphemistically referred to as the "chipshock") sent production costs spiraling upward just as the Japanese economy entered the 1980-81 recession period, putting the industry in a cost-price squeeze and some companies at the point of bankruptcy.

In the last half of 1980 and for all of 1981, the average operating rate for the Japanese paper manufacturers did not rise above 90 percent, and average operating rates for board producers was close to 70 percent, definitely in the unprofitable range. The industry itself finally approached MITI and asked for administrative guidance in the form of application of the existing Structural Improvement Law to the paper industry. This law was originally passed in 1978 to assist structurally depressed industries. The law provided a variety of remedies including coordinated production restraints and legalization of recession cartels. The law was set to expire in June of 1983 and covered 14 industries among which were some segments of the pulp and paper industry.

In the spring of 1983, the Diet passed a replacement of the original law including eight industries, many of which were under provisions of the original law.

It seemed clear that prices would remain depressed as long as there was overcapacity and therefore overproduction in the Japanese pulp and paper industry. The purpose of the recession cartels was to curtail production equally, so that prices could be supported at least at marginal cost levels. The Structural Improvement Law then contained provisions for companies to work together to reduce obsolete capacity selectively and to move the industry toward products which are higher in value-added and thus more congruent with Japan's industrial constraints and dependency on imported raw materials.

The approved recession cartels affected high-grade printing papers, coated stock, and kraft papers. For the printing and coated stocks, cartels were in effect from May 1981 to August 1981, and later extended twice to February 1982. The kraft manufacturers participated in a cartel from June 1981 to December 1981. The three cartels cut production of their products by 50 percent of total capacity. Even this cooperative situation failed to do more than support existing prices, as the Japanese economy, and therefore paper demand, failed to recover.

The SIL also approved tie-in arrangements between competing companies in buying and selling, as well as encouraging exports by excluding export volume from the production restraints. It also provided a forum in which the paper companies could freely discuss equitable methods for reducing the least efficient capacity without violating Japan's antitrust law.

Another result of the SIL was to limit investment in new capacity. A group of industry executives indicated to MITI that 80 percent of new capacity plans would be discarded and only replacement of existing facilities would be considered. The 17 paper companies represented in this group comprised 86 percent of all Japanese paper capacity. Their goal was to reduce their capacity by 11.6 percent, or nearly 11 percent of total Japanese paper capacity. The law required approval of at least two-thirds of all makers for any capacity reduction.

The situation in paperboard, however, was not so cooperative. The deadline for implementation of the cutback plan was the end of 1983, and the top paperboard makers could not agree at that time to any plan. Opposition was voiced from the beginning by Settsu Paperboard, the largest manufacturer of paperboard in Japan. Chuo, another large manufacturer, actually announced plans to add capacity by rebuilding one machine and adding another new one. With such reluctance on the part of market leaders, it is unlikely that other producers will reach agreement.

While there were some calls to denationalize industries in a number of countries including Chile and Brazil, most governments continued to subsidize to some degree their pulp and paper industries rather than let them fall prey to the combined shocks of international competition and economic recessions. Exactly what success any government's policies can effect for its industry is still to be determined. Just how much or for how long national governments will continue to subsidize or otherwise assist an industry attempting to readjust is also yet to be seen.

International Trade Issues

Similar to the increased government interest in subsidizing and/or restructuring the pulp and paper industry, international trade policies and regulations became relatively more important after the 1975-76 recession. This is not coincidental. Threatened national industries were more critical of alleged and sometimes proven "unfair" advantages foreign competitors were receiving and as a result were more keen on acquiring trade protectionist measures from their own governments. Because the greatest threats faced EC industries, much of the controversy centered on that contingent although it often involved North American and Scandinavian competitors as well.

Tariffs and Quotas

Perhaps the major issue involving tariffs and quotas, and a good example of how trade agreements came under pressure following the mid-1970's recession, was the evolving tariff and quota relationship of the EC and the European Free Trade Association (EFTA). EFTA included such major competitor nations in pulp and paper as Sweden, Norway, and Finland. In 1972, when the pulp and paper industry was experiencing a virtual seller's market, the EC and EFTA nations signed an agreement which began in 1973 to gradually increase quotas under preferential tariffs while gradually lowering tariffs as well until tariff barriers were to be abolished on January 1, 1984. In separate agreements with Sweden, Norway, Finland, and Austria, the EC agreed to begin preferential tariffs at 10.5 percent compared to its full external tariffs of 12 percent (applicable to all other non-EC countries). Ceilings were set for different pulp and paper products over which the full tariff would apply. Based on the assumption that EC consumption of these products would grow at 5 percent through the decade, EFTA ceilings were set to rise automatically 5 percent a year as well. A provision in the agreement, however, stated that these ceilings could be frozen if the EC experienced "serious economic conditions."[32]

By 1975, the earlier growth predictions had been seriously revised, and EFTA imports into the EC had become a sensitive matter. By year-end, EC authorities acted, freezing import ceilings for 1976 at their 1975 levels. EFTA exporters were particularly concerned at the compound effect this had, especially as the differential between their preferential duty rate and that of the ordinary duty rate became progressively larger. For 1978, the EC Commission again recommended freezes on eight out of 18 categories of pulp and paper imports. In March 1978, the EC presented a compromise, with five ceilings remaining frozen for the upcoming year.

Despite continuing pressure from national industries within the EC, the EC Commission did not want to undermine the spirit of the earlier agreement with EFTA aimed at sponsoring free trade in Europe. It resisted freezes in 1977 as well as demands from Cepac, the EC-wide paper industry organization, for further freezes in 1981. Surprisingly, Cepac's rationale for asking for these freezes was not the unfair advantage of state subsidies in EFTA nations. Furthermore, a clause in the original agreement allowed for a revision on the ceiling growth rate to be redetermined in mid-1977. The Swedes had originally asked for this clause in 1972 when they projected that the EC's growth in consumption might exceed 5 percent. As mid-1977 approached, top-level industry leaders from EFTA met with authorities in Brussels. Sweden and Austria pushed for a revision to 7.5 to 8 percent annual increase of ceilings, while a number of EC member countries pushed for a decrease to 2.5 to 3 percent growth. In the end, the problems involved with negotiating a change loomed greater than insisting on one, and the status quo prevailed.

In the EC-EFTA controversy on tariffs and quotas, the UK was caught in the middle. When the UK was a member of EFTA in the 1960's, the new competition from Nordic states in particular was credited for the loss of approximately 20 UK pulp and paper companies. When the UK joined the EC in 1973 it agreed that it would be unfair to former EFTA partners if the country immediately adopted the EC's tariff barriers against them. Instead, it was agreed that the UK tariffs toward EFTA would rise from zero in 1973 to 8 percent in mid-July 1977 when they would meet the progressively decreasing EC tariffs toward EFTA. As a result of the UK joining the EC and as a result of the country's subsequently awkward tariff position, the UK paper industry was particularly hard hit. During its first five years in the EC, UK paper and board exports to EC member countries fell 62 percent and imports rose 78 percent.[33]

Yet despite the EC's general reneging on its early 1970's agreements to grant preferential treatment to imports from EFTA, on January 1, 1984, the EC granted duty-free access to the EC for EFTA newsprint exports. The thriving publishing industries in several EC nations saw it to their advantage to gain greater access to lower-cost newsprint available from EFTA producers than that available from EC producers. While a favorable development for EFTA producers, this change in policy was not viewed favorably by the Canadian industry. First of all, Canadian newsprint exports to the US were already being affected negatively by the emergence of growing southern US production, reducing and disrupting Canada's export markets in the eastern US. To offset this decline, the Canadian industry hoped to increase newsprint exports to the EC. Canadians exported 1.5 million tons of newsprint duty free to the EC in 1983 and were afraid of the impact of duty-free newsprint arriving from EFTA nations on Canada's own duty-free quota, believing that Europe might be creating a trading bloc to the attempted exclusion of North Americans. Supporting the continuance of the duty-free quotas for Canadian newsprint were the major publishers in the EC who did not wish to see discrimination against suppliers leading to less competition and higher prices.

Complaints Concerning Dumping

Related to the issues of tariffs is the issue of dumping, since increased tariffs can be applied or at least threatened as retaliation against dumping. Again, the impetus to charge and subsequently investigate situations of dumping appears to have been closely tied to how threatened a particular industry or company felt itself to be. As with any issue involving one competitor requesting a government to demand another competitor to raise "unfair" prices, there are those who stand to win and lose by such an action--even within the same country. Governments, therefore, can find themselves caught between conflicting interests.

Major dumping investigations arose subsequent to the 1975-76 recession and were spearheaded by France, one of the most ailing national industries. In 1977, the French Pulp and Paper Confederation, responsible for about 75 percent of the kraftliner produced in the EC, lodged a complaint with the EC against US kraftliner producers for dumping their product in Europe and injuring the EC industry. The complaint was supported by the paper industries of most member countries. The charge claimed that US producers had not adequately accounted for transportation and handling costs in their prices quoted to European buyers. According to French estimates, if these costs were considered, then US producers were selling in Europe at prices $40/ton below domestic prices in the US. The French asked for a compensatory duty (in addition to the current 8.5 percent duty) of $40/ton. After investigating the charge, the EC Commission imposed an antidumping duty on kraftliner imported from the US beginning in March 1978. In addition, it received the voluntary consent of five other supplier countries--Sweden, Finland, Portugal, Austria, and Canada--to respect the same conditions. In exchange, investigations concerning these countries were dropped.[34]

Other antidumping efforts undertaken by the French were not as easily or quickly settled in their favor as this first one. The probable reason for this was that other interested parties within the EC and sometimes within France itself opposed these efforts.

In January 1978, the French Federation of Pulp Producers lodged a complaint with the EC Commission against US and Canadian pulp producers, claiming that they were selling pulp in Europe at dumping prices which did not take into account transportation and insurance costs of $42 per ton. The federation was supported by the French and Belgium governments who wanted a compensatory duty similar to the one awarded in the kraftliner case. In the meantime, the French government required that pulp exporters obtain a special import license, thus putting the government in a position to quickly know the source and movements of pulp into France.[35]

The complaint was quickly criticized, however, by the Comptoir des Matieres Premieres de la Papeterie (CMPP), Europe's largest pulp buyer and also the buyer for many of France's nonintegrated paper mills. The CMPP challenged the assumptions behind the complaint and also questioned why the North Americans were being singled out when pulp prices were the same in Europe irrespective of source. The British Paper and Board Federation, representing a national industry which was a proportionately large importer of pulp, joined the CMPP in its counter-complaint.[36] The investigation of the pulp dumping charge ran on for years, an apparently insolvable issue.[37]

Similarly, in 1982 the French government again attempted to establish a case of dumping against US kraftliner exporters. However this time independent box makers in France and other EC member countries opposed the attempt, predicting mass closures if the new prices were adopted. The issue became more and more split along national lines depending on the degree of integration (kraftliner-boxes) which existed within each national industry. Supporting it were France, Holland, and the UK. Opposing it were the Federal Republic of Germany and Italy. During the year, EC officials went on fact-finding visits to the US, Canada, Sweden, Finland, Austria, Portugal, and the USSR. In early 1983, they decided to act favorably on the French request and set higher minimum prices for kraftliner in the EC.[38]

Thus, raising dumping complaints like raising tariffs (or not lowering them) and not raising quotas were similar reactions of threatened national industries. As with all issues of this sort, a single "national interest" is hard to define, since some local parties stood to lose by the moves aimed at assisting their compatriots. Still, the general threatened climate prevailing through most of the EC industry as a whole caused them to move in the direction of a more fortress mentality.

International Price Fixing

Accusations of price fixing are endemic to the pulp and paper industry in the US (refer to Exhibit 4-1). However, beginning in the late 1970's and continuing into the 1980's, international price fixing came under investigation as the EC Commission continued to deal with its new laws regulating competition in and among its member countries.[39] The procedures for investigating price-fixing charges were long, and years could elapse before a decision was reached. If a decision was made against participating parties, the companies involved could be fined up to 10 percent of their previous year's sales.

For example, an investigation began in 1977 when it was alleged that members of the Genuine Vegetable Parchment Association informed each other of prices and price changes and thus determined price increases for EC markets. A decision was brought against the accused parties in December 1979, and six EC paper companies were fined. Also, in early 1977 the EC began an investigation of a possible newsprint cartel. Suspicions had been aroused when Scandinavian and Canadian suppliers announced identical prices in early spring. In addition, currency exchange clauses for UK imports to cover possible pound sterling devaluations were found to be very similar.[40] In 1980, Swedes and Norwegians were told that their selling through a joint sales agency to the UK was unacceptable.[41] It is interesting to note, however, that local newsprint producers smaller than those in North America and Scandinavia--for example those in the UK, France, and Italy--stood to lose by these EC actions and the consequent potential drop in newsprint prices.

In 1977, the EC Commission also began looking into a possible pulp cartel. In September 1981, 57 organizations from the US, Canada, Finland,

Norway, Sweden, Portugal, Spain, and the UK were sent a "statement of objections" claiming that they had fixed bleached kraft pulp prices in the EC between late 1974 and early 1978 in order to avoid drops in prices. The case was apparently based on direct evidence such as letters and memos suggesting collusion and indirect evidence such as the occurrence of simultaneous and uniform price announcements.[42] In April 1981, the Commission offered a unique proposal to resolve the case. The proposal called for the end of quoting of market pulp in US dollars, thus ending what the DG4 saw as "publicizing" pricing. The industry reaction was generally negative. One spokesman claimed this new system would only cause confusion and that pulp buyers would need to employ currency traders--an added expense weighing more heavily on the small than the large.[43]

Despite the relative newness of anti-price-fixing procedures in Europe, they will be a reality that the international pulp and paper industry will have to contend with both now and in the future.

Exchange Rate Management

A national currency that is strong in relation to the currencies of other competitor nations is viewed as a disadvantage to the nation's international trade and consequently a national industry's global competitive position. Throughout the period covered by the study, complaints concerning the strength of their national currencies are heard from various countries including the US, Britain, and Japan. A national currency that is relatively weak or recently devalued, on the other hand, should work to the industry's advantage, typically increasing exports and decreasing imports. In fact, the industry can take advantage of a devaluation of its currency against the currency or currencies in which its exports are priced to do one of two things. It can attempt to keep international prices steady, thus increasing its revenues and profits in local currency, or it can take advantage of its new cost advantage to lower prices and attempt to gain market share. The ability to successfully follow one strategy or another is not entirely in the hands of a national industry, however. It depends as well on the reaction of the importing nations.

The most controversial issue involving national currency devaluations in relation to the international pulp and paper industry were the Scandinavian devaluations in the late 1970's and early 1980's. Forest product exports of Sweden, Norway, and Finland are proportionately high. For example, they account for approximately 20 percent of all Swedish exports. Thus devaluations in these currencies are sometimes viewed by competitors in this industry as predatory moves. Also, since these countries target very similar export markets, they watch each other's currency devaluations closely and move to a significant degree together, although other national economic considerations account for the fact that the devaluations of the three currencies are not the same across the board in terms of timing and magnitude.[44]

The Scandinavian devaluations in the late 1970's, combined as they were with ailing industries in many EC member countries, no doubt added to the controversy surrounding the reevaluation of the EC-EFTA tariff agreement. However, it was the Swedish devaluation of 16 percent in late 1982 that invoked the harshest response. At the time, Scandinavia claimed that the devaluation would not be used as a measure to lower prices in the EC in an attempt to gain increased sales. Instead, the added revenues from current sales at current US dollar-denominated prices were needed to offset higher operating costs, particularly higher fuel and transport costs.[45] The industry also claimed that they feared EC countermeasures if they did indeed lower prices. However, the devaluation came at a time of falling demand in Europe and at a time of overcapacity in both Europe and the US. Prices in newsprint in the US had already begun to drop, set off by moves by major corporations there. And in the EC, buyers of various pulp and paper products pressured Swedish suppliers for lower prices. Despite the fact that there were as yet no signs that the Swedes were lowering prices, competitors within the EC became nervous. By year end, they had succeeded in rallying EC action against the Swedes' "unfair" trading due to the krona's devaluation. As a short-term measure, the EC raised tariffs on most grades of Swedish paper from an average of about 4 percent to nearly 11 percent.

In summary, various and intertwining trade issues--tariffs, dumping charges, and currency devaluations--were of particular concern to the pulp and paper industry especially within the changing European environment. They were used both to assist one's own ailing industry and as counterattacks on other nations' attempts to assist their industries.

CONCLUSIONS AND IMPLICATIONS FOR COMPETITIVENESS

Many national policies and regulations can affect the global competitiveness of a national industry. Certain conclusions concerning these effects can be applied to the international pulp and paper industry:

1. National industries may perceive that certain government regulations are unique to their nation and put them at a competitive disadvantage when in fact competitors in other nations face similar legislation, e.g., the US pulp and paper industry's concern with antipollution legislation.

2. Similarly, a national industry can focus on one aspect of government policy (e.g., antipollution measures in the US) and fail to see that other measures help balance these (energy policies in the US that kept energy costs significantly below those in Europe).

3. The more secure a national industry feels, the less vigorously it will push its government for competitive incentives and the less likely it will be to push for protectionist measures against other national industries, e.g., the US pulp and paper industry.

4. The more threatened a national industry is, the more likely its government will attempt to interfere on its behalf, e.g., the pulp and paper industry in France.

5. The more important the industry is to the national economy as a whole, the more likely its government will interfere on its

behalf, e.g., the pulp and paper industries in Scandinavia and Finland.

6. Similarly, the less important the industry is to the national economy, the less likely its government will interfere on its behalf, e.g., the pulp and paper industry in Britain.

7. National governments or intergovernment associations must balance the needs and requests of many interest groups, e.g., the dumping issues in France and the EC.

8. If a government of a major competitor industry interferes on behalf of its industry, industries in other countries will attempt to rally their own governments for measures to offset the original interference, e.g., the increased EC tariff to offset the Swedish devaluation in 1982. Their success, of course, will depend on how they measure up on issues noted in conclusions 4-7.

As stated earlier, governments can and have interfered on behalf of their national industry. However, their interference is often circumspect. To be sure, the Scandinavians have much at stake in their pulp and paper industries and would like to gain particularly in the EC and particularly at the expense of North American competitors. To what degree they can accomplish this in the face of fearful EC competition and the natural competitive advantages of the US industry is yet to be seen.

Also, to what degree governments will actually support ailing industries and keep them artificially competitive is also uncertain. Governments balk at seeing industries fail. On the other hand, some early reactions to the post-oil-price-hike environment have been reconsidered. The initial popularity of major government interference evident in the late 1970's in nations as far flung as France, Canada, and Japan appears to have waned. Most government policies that affect global competitiveness in this industry appear to be ad hoc and largely aimed at keeping the status quo, which has been under pressure since the mid-1970's, from changing too quickly.

Footnotes

1. See Toyne, Brian, *et al.*, *The US Textile Mill Products Industry* (Columbia, SC: University of South Carolina Press, 1983).

2. American Pulpwood Association, *APA Legislative Handbook* (Washington, DC: 1981), pp. 2-4.

3. Osborne, William O., *The Paper Plantation: Ralph Nader's Study Group Report on the Pulp and Paper Industry in Maine* (New York: The Viking Press, 1974).

4. "US Conferences Focus on Environment," *Pulp and Paper International* (June 1973), p. 11.

5. *Ibid*.

6. Pollitzer, Stephanie, "Capital Spending," *Pulp and Paper* (January 1983), p. 80.

7. *Ibid*.

8. "In-Plant Treatment and Closed Systems Emphasized at 15th Eucepa Conference," *Pulp and Paper International* (September 1973), p. 53.

9. *APA Legislative Handbook*, p. 5.

10. *Ibid*., p. 21.

11. *Ibid*., p. 22.

12. *Ibid*., p. 23.

13. Kalish, John, "Government Halts Industry Expansion," *Pulp and Paper International* (December 1973), p. 10.

14. *Pulp and Paper International* (December 1975), p. 5.

15. *Ibid*., (May 1980), p. 20.

16. Hagler, Robert, "How Tax Laws Affect Forest Yields," *Pulp and Paper International* (January 1982), p. 59.

17. Bayliss, Martin, "Focus on Finland," Paper 18 (April 1983), p. 20.

18. *Pulp and Paper International* (November 1980), p. 88.

19. Relatively recent US legislation (HR 4310) does provide more positive tax incentives for reforestation. This legislation, however, is aimed more toward the industrial growers and its effects will not be calculable for some time. (Hagler, "How Tax Laws Affect Forest Yields," p. 62.)

20. <u>Pulp and Paper International</u> (June 1973), p. 51.

21. <u>Ibid</u>. (August 1975), p. 12.

22. <u>Ibid</u>. (November 1981), p. 11.

23. The international economies of the industry were also affected by the increased transportation costs due to increased oil prices.

24. <u>Pulp and Paper International</u> (November 1981), p. 11.

25. <u>Pulp and Paper International</u> (June 1983), p. 5.

26. "EEC Paper Industries Facing Difficult Times," <u>IFO Digest</u> (January 1983), p. 6.

27. The weakening of the US dollars against European currencies during this period increased the competitive difficulties of Europe.

28. <u>Pulp and Paper International</u> (July 1977), p. 38.

29. In the Dutch case, Dutch competitors in fact lodged the complaint against their own government.

30. The Free Trade Agreement signed by the EC and the European Free Trade Association in 1972 stated, "Any public aid which distorts or threatens to distort competition by favoring certain undertakings or the production of certain goods, is incompatible with the proper functioning of the Agreement."

31. <u>Pulp and Paper International</u> (July 1976), p. 24.

32. <u>Ibid</u>. (1973 Review Number), p. 13.

33. <u>Ibid</u>. (June 1978), p. 86.

34. <u>Ibid</u>. (March 1978), p. 14.

35. <u>Ibid</u>.

36. <u>Ibid</u>. (April 1978), p. 10.

37. The last reference the author found to this case was in "The Back Page" of <u>Pulp and Paper International</u> (November 1982). The author, Martin Bayliss, reports, "This [case] is now several years old but is still, apparently, under active investigation, contrary to recent reports."

38. <u>Pulp and Paper International</u> (January 1983), p. 7.

39. Articles 85 and 86 of the Treaty of Rome.

40. <u>Pulp and Paper International</u> (May 1977), p. 11.

41. *Ibid*. (September 1980), p. 114.

42. *Ibid*. (November 1981), p. 10.

43. Kalish, John, "The Back Pages," *Pulp and Paper International* (June 1983).

44. For example, Norway's ownership of North Sea oil partially accounts for the fact that its currency did not tend to devalue as much as the Swedish krona. Finland, with a high foreign debt and a legal agreement with unions stating that wages would be renegotiated in the event of major devaluations, also tended to be more conservative in its devaluations. Sometimes these governments would attempt to assist their local pulp and paper industries by other means in the face of a larger Swedish devaluation. For example, when Norway disappointed its pulp and paper industry by not following a Swedish devaluation of 16 percent and a Finnish devaluation of 10 percent in late 1982, the government promised a cancellation of the electricity tax for 1983 and added investment credits.

45. The US dollar was the medium of quotation for international pulp and paper prices in Europe. If the dollar rose in value against an exporter's local currency and if the rise was unaccompanied by a drop in price, an increase in revenues in local currency resulted for the exporter--and vice versa.

EXHIBIT 4-1a

Government Concerns Appearing in U.S. Pulp and Paper Corporation's Annual Reports 1970-1982

	Environment				
	Pollution Regulation	Forest Policy	Anti-Trust	Price-Fixing	International Trade
1979-1982 (5 years)	14	4	18	13	3
1975-1978 (5 years)	5	4	12	5	-
1970-1974 (4 years)	1	-	-	-	-
Total	20	8	30	18	3

EXHIBIT 4-1b

Companies Surveyed 1970-1982
Annual Reports

Boise Cascade
Champion International
Chesapeake
Crown Zellerbach
Fort Howard
Georgia-Pacific
Great Northern Nekoosa
International Paper
James River
Kimberly-Clark
Louisiana Pacific
Mead
Potlatch
Scott Paper
Sonoco Products
Southwest Forest
Union Camp
Westvaco
Weyerhaeuser

EXHIBIT 4-2

Level of Concern Relating to Government Issues Internationally by Year

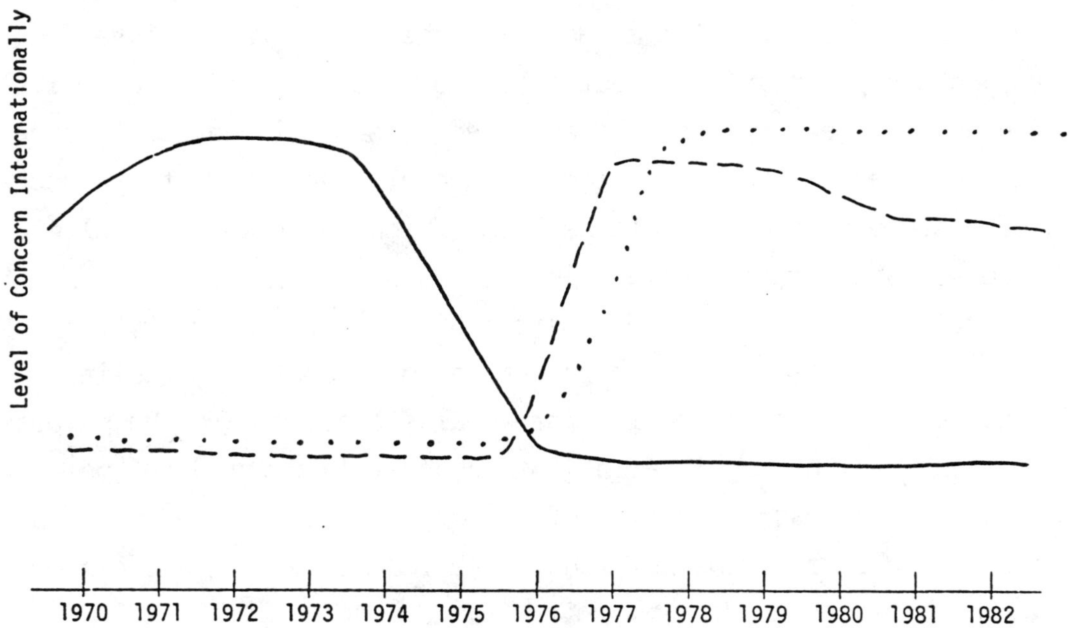

——— Environmental Policies

----- Government Subsidies

..... Trade Issues

CHAPTER 5

CORPORATE PERCEPTIONS AND COMPETITIVE STRATEGIES

The previous chapters' discussion of the emerging economic and political dimensions of the industry provides a backdrop to this chapter's discussion of competitive strategy. These trends are often reflected in the strategic planning process carried out by managers of firms in the industry, as illustrated for example by the James River Corporation of Virginia. The managers of this organization have focused strategy on the potential for restoring the value of antiquated mills and papermaking equipment that can be acquired at something close to book value and applied to the production of special paper products to meet the needs of customers in market niches. This blend of technological and market focus, while rather unique in the industry, is quite appropriate to the particular competitive dimensions of the paper industry at the present point in time. Such approaches to the economic and political circumstances of domestic and international situations are the purpose of this chapter, and the remainder of this chapter is devoted to the identification and categorization of the strategies employed by firms in the various segments of the pulp and paper industry.

RESEARCH METHODOLOGY

When this study was initiated it was decided that the work of Michael Porter would provide the analytical framework for the study. Porter's theory suggests that five forces operate to shape the competitive environment of an industry or segment of an industry. These are the ease of exit and entry into the industry, the relative power of suppliers of the industry, the degree of threat from substitutes for the products of the industry, the relative power of the customers of the industry, and the nature of the rivalry among firms in the industry, recognizing that the latter may be shaped in part by the other four factors. Porter goes on to suggest that, given the competitive nature of the industry, firms within the industry may chose to compete in three different fashions, employing what he calls overall low-cost strategies, differentiation strategies, or strategies that concentrate on one of the above strategic ploys in respect to special groups of customers.

The researchers who were involved in this portion of the study collected information about these strategic dimensions of the industry. They began with accounts of the strategies used by firms and segments of the industry written by other observers or participants in the industry itself. Attempts were made to see whether these accounts squared with secondary data about the industry, as well as with information obtained in corporate annual reports. Additional information about competitive strategies emanated from two principal sources. The first of these consisted of structured discussions of strategy in the pulp and paper industry with selected top-level managers in various components of the industry who agreed to be interviewed. The researchers asked direct questions derived from Porter's model and took notes on the replies and elaborations to the questions. These interviews were conducted at the site of the firm and usually led to the provision of various documents that would help the researchers understand the nature of the firm and its strategy. Often the interview was followed by telephone conversations to gain additional information that became important as the interviews progressed. As indicated, these interviews touched most segments of the paper industry and included visits with managers in selected firms and associations in England, France, Germany, Finland, Canada, and Japan, as well as with managers in a host of firms in the United States.

The second source of information about competitive strategy in the pulp and paper industry was a survey of firms in the United States using a questionnaire based on Porter's concepts. This questionnaire was expanded to also solicit information of a more general nature, along with information about foreign trade issues related to the global dimensions of competitive strategy. Three hundred sixty-three firms and business units of firms responded to this questionnaire. The details of these responses are provided in the Statistical Addendum, which is available from the Division of Research, College of Business Administration, University of South Carolina.

Each of these sources is drawn on in the sections of the chapter which follow. No attempt is made to attribute the source for each comment or observation about the competitive structure and competition in the pulp and paper industry as these tended to meld in the analytical process associated with the development of the presentation.

One problem that will be encountered in the following discussion is the classification of the groups of firms that make up the paper industry. Porter does not help us solve this problem of industry boundary identification, and in his own work he establishes the working definition of an industry as the group of firms producing products that are close substitutes for each other, while he acknowledges the controversy that surrounds this definition. Much of the information used in this study had been collected within this same framework, so the other portions of the study adopted the same working definition. The term "working definition" is used by both Porter and the present researchers because it is hoped that further analysis of the data collected in industry studies will reveal "strategic" rather than product groupings. Such a novel grouping would emphasize the comparative advantages that firms hold in an industry that permit them to profitably distinguish themselves, at least in the intermediate and short run.

OVERVIEW

It is clear that the paper industry is not a special case in respect to the nature of competitive strategy. All the themes present in other industries were found to be a part of the strategic thinking of managers in this important element of the manufacturing sector of the US economy and of the economies of most other nations. The industry is made up of distinct elements as already noted in the previous chapters, but many of the firms overlap these segments in their vertical and horizontal structures. It was evident that managers on many occasions were ready to exclude certain firms from the industry because of differences in markets, technologies, and strategies. This is largely the result of the process orientation of particular managers and firms. Those firms closer to the trees tend to have managers who think in terms of forest products and paper as but one among a number of options, while converters deal in a number of raw materials of which paper is only one. When the survey questionnaire was mailed to the 1,385 firms that the Ward organization identified as declared members of the paper industry in 1979, a large number were returned unanswered because the recipients no longer considered their firm to be a part of the industry. In most cases their tenuous link with the industry had been severed when plastics replaced paper as a raw material for their firm. Others decided in the four-year interim since the publication of the Ward directory that the message was the medium, and they chose to classify themselves as printers or graphic artists rather than converters of paper products. While no direct replies to our queries indicated that a forest resources firm had divorced itself from the paper industry, this possibility also existed.

There is no question that imaginary boundaries exist in the minds of managers beyond which a firm or unit is outside the industry. The interviews revealed that the pale beyond which a "pulp" or "paper" firm does not go is mixed in the minds of managers. Many firms feel constrained by this boundary in the strategy formulation process as they consider their competitive talents limited by the industry. This comment is made early in the discussion of competitive strategy in the industry because theory does not place such limits on strategic position. Perhaps the theory of strategy formulation should recognize such semantic limitations on the scope of strategy consideration.

This problem of industry identification confounded the study. But all the competitive themes were discovered, and these were generally dispersed throughout the somewhat arbitrary segments of the industry used to categorize the firms. A review of annual reports from many well known manufacturers provided citations of the strategic importance of numerous competitive strategy themes. The following were among the more prominently mentioned ones:

1. Achieving market leadership
2. Innovating and developing new technologies
3. Identifying and focusing on new markets
4. Achieving cost reductions and redeploying assets in more profitable and diversified ventures
5. Emphasizing product quality and shifting to higher value-added products
6. Achieving more rigid cost control and maximizing productivity
7. Gaining access to capital at a reasonable cost
8. Developing raw material self-sufficiency
9. Reducing dependence on domestic markets by exporting
10. Shifting to consumer products and deemphasizing the commodity nature of products.

These were common themes but somewhat indiscriminately mentioned, without regard to their sometimes mutual exclusiveness. The same result was obtained in the interviews and in response to questions that probed these themes in the questionnaire.

Whatever the true dimension and boundary of the industry, there was consensus that the overall industry has hurdled the strategic barriers placed in its path over time to remain one of the most viable of manufacturing industries. The industry overcame the initial technological barrier of hand manufacture to become among the first continuous process manufacturers. It went on to solve the raw material problem that limited the full exploitation of this form of technology and linked itself with the forest products industry at a point in time when one of the major uses of wood was being displaced by other sources of energy. Both of these strategic innovations were driven by the information revolution that the industry was then able to exploit to gain the utmost advantage from such innovations. In the process the industry created new raw materials that became substitutes for many other products and the basis for the creation of other industries.

If there was any breakdown in the chain of strategic successes, it may have been in the loose connection between what was then known as the paper industry and these satellite developments. One cannot fault the industry for not integrating the printing applications that resulted from these evolutions because the technology of printing was already well developed. However, many of the new applications of paper as a raw material were derivatives of papermaking technology itself. One can understand the strong craft orientation to paper as the product, however, and this fact still drives much of the strategic process and explains the alliance that exists between the manufacturers of papermaking machinery and the papermaking industry. This issue of culture as a strategic constraint is not unique to this industry, but it is more apparent than in industries where technological achievement explains less of the strategic history of the industry.

Strategic Imperatives

Throughout the course of the study there was an obvious tension expressed about the strategies of concentration and integration. The papermakers prefer to have their firms grow and earn profits from the production of paper, yet they recognize that the profits in papermaking accrue as rents to the forest land holder. The treecutters have the same complaint, but they are often able to exact a profit from paper in the dynamic interplay between the uses to which their product can be applied. They also have a natural advantage in the often by-product nature of the material supplied to papermakers that permit sales to be calculated as almost pure profit. At the other extreme the marketers of paper products gain the classic profits derived from trucking and barter.

Conversion of paper is an alternative for the papermaker, but the value added is offset by the costs of production unless the manufacturer is willing to exploit the lower-cost labor that is feasible with most converter technologies. If papermakers wish to integrate forward to the ultimate marketplace, they must consider the conversion opportunities, and many have. Usually the integration is into lines of products that have similar technological characteristics to papermaking itself, but some firms obviously have been seduced by the lure of profits to abandon the full expression of the traditional culture in papermaking. Perhaps some of the firms that provide this impression are the result of backward integration toward papermaking--converters and merchants who were not steeped in the technology but found some historic advantage in gaining control of supply. The present dilemma of boxmakers in respect to the rapidly rising cost of linerboard may provide an example of why firms made such strategic decisions to enter and attempt to shape a prevailing culture. This latter example provides a chance to suggest that the dilemma of the boxmakers may in part be a function of the alternatives the paperboard makers have to become profitable through the exercise of comparative advantage in foreign markets.

In contrast to vertical integration, there was much less discussion and evidence of horizontal integration and diversification among the industry's firms studied, although many of these firms indicated that this had been a part of past strategies. The times in general seem to favor market concentration, and many firms in the industry as a whole are following the trend toward divestiture of unrelated lines of business. The search for market niches that provide insulation from rampant competition has led to increased specialization of product lines. This also has led some firms to accept residual positions in declining portions of the industry where profits can be eked out by the essential absence of competition. Undoubtedly such end-game strategies are being studied by other firms in the industry, although this was not a business strategy topic that many firms wished to discuss.

COMPETITIVE STRUCTURE OF THE INDUSTRY

It has been argued that the essential understanding of competitive strategies used by firms in an industry depends on an appreciation of the relationship between such strategies and the environment in which the firms competitively exist. Firms develop strategy from the assessment of the forces that impinge on the industry and shape the strategies of competitors but respond differently to the shared environment because of competitive advantages that make them unique.

Porter states that competition in an industry is rooted in its underlying economic structure, and this structure depends on five basic competitive forces. These are the threat of entry into the industry by new firms, threats of substitutions, the bargaining power of suppliers, the bargaining power of customers, and the rivalry among current competitors. Each of these forces provided a category for the collection of information in this study of the industry, and the sections that follow are divided accordingly. These results are then followed by discussion of the actual strategies used by firms to deal with competition in a profitable fashion.

Barriers to Entry and Exit

The threat of entry by new firms into the industry obviously has profound impact on the competitive structure. New entrants bring new capacity to the industry, they disturb the competitive rivalry through their desire to gain market share, and they vie for the raw materials and other resources required to do business. Industries can be categorized by the barriers to entry that prohibit new competition. The following are among the more important entry barriers:

1. Economies of scale required to do business in the industry
2. The opportunities for product differentiation by established firms that commands customer loyalty and brand identification

3. Capital requirements that may provide no guarantees of success

4. The cost that customers might have to incur to switch to a new supplier

5. The need to secure channels of distribution that might not be available to outsiders

6. Other unique advantages such as proprietary products or work sites commanded by present firms that cannot be easily duplicated

7. The barriers erected by government that limit or ease access to an industry.

It is clear that new entrants in the primary products sector of the paper industry face extensive capital cost to participate, and the cost of capital was such as to limit the potential profit that normally could be obtained from such investments. The degree of vertical integration that is most economical in the primary products sector provides for minimal efficient scales of operation that have a cost close to $500 million. In addition, most sites in the world suitable for integrated pulp and paper mill operations which command woodlands of sufficient size to provide the necessary feedstock have been occupied. New entrants also face extensive existing capacity, although entrance could be through the acquisition and conversion of some portion of this existing capacity. In fact this may well be the only profitable way in which firms can continue to enter the primary sectors of the industry, given the present and predicted rates of growth in sales.

While new greenfield sites for primary paper production may develop in nations with extensive afforestation, such developments will require considerable additional capital investment to provide the infrastructure to support such heavy industry. Such developments would also require entry into existing markets or the development of domestic demand for these products. The economies of scale that now exist present a serious question whether initial domestic demand would be sufficient to support such capacity operations, leaving the industry dependent on exports into existing markets. The Scandinavian countries provide an example of an emergent strategy of market penetration through a roll-out from pulp export to the export of primary paper products with more extensive value added. However, the viability of this strategy is still in question, and the move toward joint ventures between Scandinavian forest products firms and other European paper manufactures to permit integrated paper production may suggest that the problems of transportation cost pose considerable problems. The US firms that responded to the survey indicated that the costs of transportation are the greatest barrier to export after the high value of the dollar is taken into account.

Less extensive are the barriers to entry into segments of the industry beyond the primary products sector. This is evident in the lower level of concentration of firms in these sectors, with the exception of the firms that manufacture sanitary paper products. In general, there is a high rate of turnover of new firms appearing to produce products derived from primary paper products. Capital costs are considerably lower in the conversion and packaging segments of the industry, and this is evident in the lower ratio of capital to labor in the production process of these firms. Firms in these sectors of the paper industry have much less difficulty exiting from the industry or a particular sector of the industry and entering new lines of business. Plant conversion is a more feasible strategic alternative in these sectors, although primary product producers have demonstrated more flexibility in the recent past. In some respects these strategies of reconstruction and rebuilding to enter into more profitable sectors of the industry are the dominant characteristic of present day industry strategy.

The sanitary paper products segment of the industry is generally recognized as the one sector that has diminished the threat of entry by product differentiation, although there is a general appreciation that product differentiation can aid profitability. The premier defense of product differentiation eroded somewhat during this past recession, as demonstrated by the gains of generic products in the hearts and minds of consumers. This battle is not finally resolved, however, and it remains to be seen whether the increasing advertising expenditures can reverse this trend. For example, new attempts to gain market share through advertising have been launched outside this sector by the Fort Howard company. Recent ads have attempted to extend consumer loyalty to some sanitary food container products as well as converted tissue products; however, the overall industry generally has the lowest regard for advertising as a competitive method, as confirmed by our survey of the industry.

Few forms of unique advantages that might provide barriers to entry exist in the industry beside those already noted in respect to access to greenfield sites. Most products and processes are not proprietary and new equipment has become available internationally to all firms as it is developed. Ownership of patents or other proprietary knowledge was generally considered to be a relatively unimportant competitive factor among those firms responding to the survey. Up until recently, firms in the pressure-sensitive tape sector held some advantage in proprietary equipment, but the equipment supply sector has responded to make such equipment available to any new entrant. Certain firms in West Germany claim proprietary skills and knowledge in the fine paper sectors of the industry that permit a worldwide monopoly in export. There may be other examples of such unique barriers elsewhere in the industry, and the potential for such advantages always exists where loyalty to the firm is a strong cultural element.

The major government policy issue that has influenced entry has been the increased capital cost associated with environment protection legislation. This cost is part of the past experience of US companies, but new entrants would view this as an initial barrier. European firms experience this as an emerging barrier because of their later emphasis on environmental factors in paper production. In addition, some governmental product standards exist in the US and abroad that can act as barriers to entry in domestic and foreign markets. These are not considered to be unsurmountable and provide more of an annoyance to firms considering foreign trade. The one instance where government policy offered a substantial barrier to entry was in respect to the French industry, where subsidization of moribund firms precluded any new entry by more efficient competitors.

Findings from the Survey

The firms in the US that responded to the survey held moderate belief about the influence of economies of scale on profitability in the industry. Firms in the various sectors must have approached this question in a different fashion, but they came to similar conclusions. Firms in the primary products sector experience high capital costs to gain economies and would be well protected from intrusion. Apparently the cost of the barrier offsets any profit advantage for these firms. The converters and packagers that are less protected from outside entry because of the lower minimal efficient scale of operation and lower capital costs do not perceive this as the cause of lower profitability or an opportunity to enhance profitability. However, the design of the question that was asked may have influenced the response as it simply asked whether difficulties in obtaining economies of scale in production restrained industry profits.

A related issue concerning barriers to entry was whether the lack of product differentiation to hold customer loyalty acted to restrain industry profitability. This was more highly regarded as a factor restraining profitability, but there was no significant difference in the response of firms in the various sectors of the industry. This simply supports the observation that few sectors of the industry have attained a degree of product differentiation that could influence profits. Apparently managers in all segments of the industry realize that profitability would be enhanced if customer loyalty could be gained, but they acknowledge the difficulty bringing this about because of the general commodity nature of the products produced by the industry.

As to whether new entries had lowered profits, firms in various sectors of the industry answered differently, and national differences in perceptions also varied. For US firms, the overall response was a moderate belief that new domestic competition had restrained profits, but a much lower belief that new foreign competition had done likewise. For example, 59 percent of the US respondents reported that new foreign competitors were not at all important in explaining restrained profits, while only 20 percent of the respondents provided a similar response to the question of new domestic competition.

Exhibits 5-1 and 5-2 illustrate the differences in response of some selected segments of the US industry to these two questions. Exhibit 5-1 shows that the paperboard manufacturers believed new domestic competition is a relatively unimportant factor in explaining industry profits, while the manufacturers of stationery were much more likely to perceive such new competition as an extremely important factor. Other manufacturers have a variety of beliefs about the importance of domestic competition, and the lack of consensus is quite significant.

Exhibit 5-2 reveals the lesser belief in the importance of entrance of foreign competitors in the US markets as an explanation of profitability in segments of the paper industry. It also reveals that a consensus does not exist among sectors of the overall paper industry. Again, the manufacturers of stationery are much more likely to indicate an impact of foreign entrants on profitability, although the degree of belief is half that expressed for new domestic competition. The manufacturers of envelopes are much more likely to see the relative unimportance of such competition on profits, followed by the manufacturers of paperboard.

For pulp and paper firms in other countries, however, foreign competition looms as a much more important factor influencing profitability. With the notable exceptions of Canada and Scandinavia with their large export surpluses and highly internationally competitive industries due in large part to their large timber resources, most other countries face significant import competition. The competition is most fierce from North American and Scandinavian firms, but is also strong from other nearby countries with similar but slightly different costs and industry structure, such as within the European Community. In fact, for the EC industry, foreign competition appears to have exerted a greater downward force on profitability than domestic competition, although competition from the former increases competitive pressures from the latter as all domestic firms lower prices and become more aggressive in response to foreign competition. Thus the less internationally competitive a national industry is, the greater the impact foreign competition will have on domestic profits.

Thus barriers to entry are important factors shaping competitive strategy formulation in the pulp and paper industry, but this conclusion is relative to the sector of the industry. As we shall see, the converters and packagers seek protection from the relative ease of entry through focus on regional markets. The manufacturers of primary paper products are much less likely to seek this form of protection because of the natural barrier provided by the high capital costs of efficient plant operations and the shortage of suitable forest lands. While the benefits of product differentiation are generally appreciated, these are less accessible to firms in the paper industry as it is quite difficult to convince the sophisticated customers of the industry that the product sold is different from the product sold by competitors.

As for exit barriers, there are many influencing factors, among the most important of which is the cost of exit. At the level of the firm, the size and value of the assets owned can represent a sizable exit cost of alternative uses are not available or no one outside the firm is willing to purchase them at their market value. In this light, the primary products sector, with its high degree of capital intensity and machinery specialization, is more costly to exit than the downstream sectors, suggesting that firms in the primary products sector will fight exit more strongly. Also, the fact that in order to acquire these sizable assets most firms have used external financing suggests that their creditors may be more willing to assist the producers in order to protect their loans. This appeared to be the case with several large and financially troubled Japanese firms in the early 1980's.

Another key consideration is the government's willingness to ease, delay, or even prevent the exit of a firm in the industry. As noted in Chapter 4, governments are more likely to follow such policies when the firm is large and the industry as a whole is an important part of the economy in terms of employment, exports, or strategic importance to downstream or related industries. The protectionist policies of most EC countries to their troubled paper industries is an illustrative example.

Buyer Power

One of the most important decisions made by a firm is the customer it choses to serve. The choice of customer defines the market in which the firm will operate. This in turn determines the industry of which the firm will be a part and the products it will produce in the longer term. There can be no more profound strategic decision made by a firm, yet this decision is often made in a most opportunistic fashion and without conscious thought of the strategic implications.

In addition, the way in which a firm positions itself in respect to customers sets the stage and potential for forward integration. It is the customer markets that a forward integrator seeks to penetrate, hoping to gain the return on value added by the customer. Of course the firm can attempt to capture customer profits in other fashions, which leads us to the issues of buyer power. It is this aspect we now explore in respect to the pulp and paper industry.

Buyers compete with the industry by forcing down prices, bargaining for higher quality or more services, and playing competitors against each other, at the expense of industry profitability. The power of important buyer groups to effect these consequences depends on characteristics of the market situation. In particular, the power of buyers depends on whether the buyers are few in number and buy a large proportion of the output of firms in the industry. The power of a particular buyer will also depend on whether the products it purchases from the industry represent a substantial proportion of the total purchases of the buyer and whether they are critical components of the products or services produced by the customer. It also depends on whether the products are commodity in nature or specialized to a particular use. Another factor is whether the buyer faces any switching costs in moving from one supplier to another. Furthermore, if the buyer earns low profits, a greater incentive exists to seek a lower price. Buyers that have the capability to integrate backward will also be in more favorable bargaining position with suppliers. Finally, the power of the buyers over firms in the industry will depend on the amount of information they have about demand, actual market price, and supplier costs.

Based on our research, the issue of buyer power is clearly relative to the degree of integration of the firm. Many of the firms that were included in the study are customers and suppliers of other firms in the study. This provides one of the more interesting aspects of this study, but it does complicate and confound the analysis. Customers are many for the majority of commodity paper and pulp suppliers, yet customers of such firms do tend to become habitual buyers from particular firms, apparently because pulp and paper prices are public information and the purchaser has technical capabilities to evaluate the product and measure the dependability of suppliers. There is no need to shop around because the discipline of the marketplace works too well. Many in the industry believe that the degree of customer loyalty that exists is the result of the potential shortages in supply that are known to occur during boom periods and the occasional industry strike. However, most customers attempt to second source both commodity and special pulp and paper products to protect against supply breakdowns and as a check on technological developments and quality control.

Converters and container manufacturers are much more likely to be wedded to particular customers by the nature of their products. In most cases the products they supply are a small part of the customer's raw materials. This is less true for the suppliers of primary paper products and pulp, although the manufacturers of industrial papers for original equipment manufacturers often face the same situation. When paper merchants are part of the channel of distribution, their bargaining power is most apparent, resulting from the large volume of their purchases and the many customers to which they distribute in turn. This is the classic power of the merchant in every field and is not peculiar to the paper industry, although many firms were quick to mention this aspect of buyer power.

The trends most frequently mentioned included a greater concentration among the buyers of paperboard and a lesser concentration among buyers of communication and printing papers. The backward integration trend that had existed in respect to publishers and printers has diminished and there is a strong suggestion that such integration will actually regress in the future period.

Overall, one gets the impression that buyers have exercised discipline on the paper industry in the past and probably will continue to do so in the future. Customers tend to be knowledgeable about the product they buy from the industry and they often are the instigators of technological change in the product. Again we note that the paper industry devotes little of its effort to research and development, and it appears to take many of its cues for technological change from customers. Many of these customers are heavily involved in new product development, which occasions special requests for raw material support from elements of the paper industry. Some paper firms maintain technical institutes specifically for this purpose.

There were no significant differences among the segments of the industry in respect to the degree of concentration of customers. For example, the US respondents' average proportion of sales to the three largest customers was 27 percent, and it was generally agreed that it was slightly more easy than difficult for buyers to switch the source of their purchases of paper products. Some 29 percent of respondents to the survey believed it was very easy for customers to switch, while only 2 percent thought it very difficult.

There was a basic agreement that a few large-volume buyers were forcing prices down and restraining industry profits. More of the respondents believed this important than did not, producing a picture of the industry that had prices determined in a marketplace where buyers had power. There was also agreement that the customers of the paper industry were unwilling to pay for higher prices for products of purported higher quality. That is, they discounted claims that the products of a particular firm were somehow superior to the products of another.

There was disagreement in the responses among firms in components of the industry in respect to whether customers have secondary sources of supply that are readily available and whether customers threaten to produce the products supplied by firms. There was an overall high level of importance granted to the first of these factors as an influence on profitability, while the latter factor was generally discounted. For example, Exhibits 5-3 and 5-4 display the differences in response among selected sectors of the overall US industry.

Overall, the power of buyers in the paper industry would best be assessed as moderate without any indication of future concentration of power in most sectors. The threat of backward integration does not seem to be great in most sectors. Firms in the paper industry do seem to be more than usually sensitive to the buyers' need for quality products, and this seems to be a function of the industrial nature of the buyers. Few sectors sell directly to final consumers, and the intermediaries are knowledgeable about the product. We can suspect that the buyers of paper products are able to obtain substantial profit from the transfer price obtained, leading one to believe that many sectors of the paper industry would be tempted to integrate forward toward the ultimate consumer.

Supplier Power

Suppliers to an industry can exert bargaining power over firms by threatening to raise prices or reduce the quality and quantity of goods and services supplied. A supplier group is particularly powerful when the supplier group is dominated by a few companies and is more concentrated; there are few other substitute products for sale to the industry; if the industry is not an important customer of the supplier group; if the products of the supplier are important inputs to the purchaser; and if the products sold by the supplier have been differentiated from the products of its competitors or if the supplier has created switching costs for the purchaser. The power of the suppliers also is enhanced if they pose a credible threat to integrate forward and usurp the role of the purchasers of their products.

With respect to both supplier and buyer power, a tension exists which is competitive in and of itself. Both buyers and suppliers can choose to compete directly in the industry by backward or forward integration, but suppliers also serve as moderators of competition among users of the raw materials they supply and, thus, raise their products to their position of greatest productivity. In other words, firms in the paper industry compete with other firms in the industry and firms outside the industry for raw materials, and the suppliers of these materials directly manage this dynamic. A like case can be made for the role of buyers, particularly when the purchasers are merchants who facilitate the distribution of products.

The whole point involved in the assessment of supplier power is the development of understanding of feasible strategies in respect to power. These include such features as spreading purchases among suppliers and having multiple sources of supply, avoiding switching costs and encouraging new suppliers to enter the field, promoting standardization of raw materials to give them a commodity versus a proprietary character, and practicing a tapered backward integration that poses a continuing threat of more extensive integration.

The paper industry worldwide is characterized generally by an extensive concentration of raw materials supply. The principal material is fiber, paper, or paperboard in every sector of the industry, emphasizing a previous comment that elements of the paper industry have vertical relationships in addition to more obvious market competitive relationships.

In the US, managers in most segments of the industry talk about backward integration to ensure sources of supply or to capture the profitability of suppliers. Many firms have such integration as a part of their history, so a pattern of backward integration is well established in the culture of the industry. As it already has been mentioned, there appears to be a greater attractiveness to backward versus forward integration that may not be explained entirely by the prospects of either short- or long-run profits. The relative degree of concentration of buyers versus sellers is a basic aspect of this phenomenon, and a review of the input-output statistics for the overall industry as well as its segments testifies to the much higher degree of supplier concentration.

On the other hand, the principal raw materials of the firms in the industry are classed as commodities. Wood, pulp, wastepaper, paper, and paperboard are each products with prices typically common to all purchasers on a regional, national, and often a worldwide basis for particular grades and quantities. There is little attempt to brand such products in the industrial markets, although certain suppliers are perceived to have more rigid quality standards, provide more reliable service, or maintain a degree of loyalty to their customers. However, any competitive advantage that accrues to these attributes of a firm usually is emulated subsequently by competitors, so the advantage is transient. Most suppliers hope to establish long-lasting relationships with customers, and the only time this is not apparent is when overcapacity to meet domestic demand leads segments of the industry without established export markets to attempt export on a temporary basis and as an economic expedient.

The European paper firms are particularly sensitive to the power of primary product suppliers because of the international market from which a large part of their supply emanates. Backward integration is an international venture on their part unless they can rely on the development of wastepaper networks. For these various reasons the manufacturers of paper in Europe are much more sensitive to the power of suppliers than is typical in the US. It is probable that European manufacturers shall seek to overcome in whatever way possible this power of the suppliers to dictate terms of the transactions, including making the threat to integrate backward more credible in the future period. This may take the traditional form of investment in supplier firms, but may also include more actual ownership and operation of facilities in other parts of the world, including the US, where the greatest comparative advantage exists. The same scenario applies to Japanese paper manufacturers, many of whom already have invested internationally to secure raw material supplies and have made significant strides to increase the use of recycled wastepaper.

The other major supplies and services used by the paper industry include chemicals and railroad transportation, which are products and services that in most countries are outputs of industries more concentrated than the pulp and paper industry. Energy is an important input for many converters and packaging firms, but the primary product sectors use cogeneration for much of their heavy energy requirements. Another raw material that seldom appears in the statistics of the industry but is critical to the operation of many segments is clean water--this raw material, along with labor, has been an important concern at various periods in the history of the industry.

Skilled labor is a particular characteristic of the primary products sector of the industry. Wage levels in the industry are much more a characteristic of the skill than the degree of organization of the labor force. The technology of papermaking has a historically strong craft identification and this does not seem to be losing its flavor in spite of extensive computerization of production processes. The paper industry provides insight on how the computerization of a manufacturing workplace may be facilitated when a continuous process technology is dominated by a craft orientation. One European firm maintained that its backward integration was the result of the need to ensure a craft outlook in the workup of paper stock used in its main production process. Another European firm stated its suppliers advertised its purchases as evidence of the high production standards of their supplies.

The already existing high degree of concentration of suppliers suggests that further overall concentration will be less likely. There is some belief that further concentration of suppliers of paperboard will take place through merger and acquisition, with important effects for the packaging components of the industry. This issue will be addressed later. Some container manufacturers have already sought to offset such concentration by backward integration, with the acquisition of the Brown Systems operation of the Continental Group by Stone Container perhaps providing an example. On the other hand the divestiture movement that exists in the broader US economy may portend a lessening concentration of suppliers to the paper industry.

In addition to confirming that the overall level of concentration of suppliers to the paper industry was high, our research generated responses that indicated that it was somewhat more difficult for paper firms to switch among suppliers than it was difficult for the customers of paper firms to switch their source of purchases. There were no differences among firms in the various components of the industry in this respect, but the degree of concentration among the components of the industry did vary in a significant fashion. As Exhibit 5-5 indicates, the responding paperboard manufacturers reported the lowest degree of supplier concentration, while the envelope manufacturers reported the highest concentration. While there is no direct way to confirm these reports, it might be noted that the paperboard industry draws its raw materials from more diverse sources than the envelope manufacturing industry. The 1972 Input-Output Analysis performed by the US Department of Commerce shows the single greatest input of the envelope industry being paperstock, and this makes up 50 percent of all purchases. The single greatest class of purchase by the paperboard industry is products of the logging industry, and these provide for 21 percent of purchases. Undoubtedly the envelope manufacturers draw on a more limited group of suppliers of paperstock than the paperboard manufacturers draw on woodlot suppliers.

Container manufacturers also report considerable large portions of raw materials being purchased from the three largest suppliers. The Input-Output Analysis indicates that this segment of the paper industry may have the greatest dependence on any single supplier industry, with 58 percent of total raw materials in the form of paperboard mill products. Some comparable statistics from the Input-Output Analysis include the supply of 24 percent of the pulp manufacturers by the logging industry, 24 percent of the papermill industry by the pulp industry, and 33 percent of the papercoating industry by the papermill industry.

The relatively few substitutes for the raw materials used by the industry was credited with being a restraining influence on profits in the industry. Some 49 percent of the respondents to the survey believed this to be more than somewhat important in explaining profitability over the last five years. There was no significant difference among respondents in different sectors of the paper industry in this respect.

There were significant differences in the responses to the three other questions that explored the issue of supplier power. These were the importance of the profit restraining influence of a few firms controlling sources of supply, suppliers having considerable volume of sales outside the particular segment of the paper industry, and the threat suppliers present to produce products manufactured by the purchaser. These results are depicted in Exhibits 5-6, 5-7, and 5-8.

Apparently the power of suppliers is a more important influence in the shaping of the competitive environment of the paper industry than the power of customers of firms in the industry, and particularly for the industries in other countries. Influence of suppliers does vary among components of the industry, however, and this suggests that backward integration or the credible threat of such integration is more extensive in certain segments of the industry, and also generally less for non-US producers.

Substitution

Firms in any industry not only compete with one another, they also compete with firms in other industries that produce products that the consumer can use to serve the same purpose. Few products lack such substitutes, and it is largely a matter of customary practice that provides any insulation from such outside competition. Firms in an industry will do everything possible to enhance the characteristics of their products to improve performance so that relative price changes produce less volatile changes in the demand for products. Where the price and performance of substitutes are closeto those provided by an industry, the profitability of an industry is limited, as any boost in profits will produce a rapid supply response. Over time, greater dynamic changes in customary usage as the result of new technological developments can _permanently_ displace products and antiquate an entire industry.

This neglect of an important topic of concern to the paper industry is the result of focus on competitive business strategies. Industries cannot stand against the shifts in technology and other aspects of culture that ultimately produce maturity and decline, although they may attempt various forms of persuasion to hold some position in the face of such fundamental change. The solution to this dilemma is not contained in the concept of business strategy; rather it is a matter of corporate strategy which transcends the nature of an industry in which a firm might find itself at a point in time. The answer to the threat of a new substitute is the conversion of resources to a new employ either in more sheltered portions of the existing industry or in new industries where the firm might demonstrate some competitive advantage.

The competitive pressure on an industry posed by product substitutes is a much discussed topic in the paper industry, and many individuals in the industry testify to the recent displacement of paper products that had been mainstays of traditional papermaking firms. These concerns address fundamental changes that stem from both technological advances and modified tastes and preferences of the public.

The major concern in the paper industry is the threats posed to the industry itself, rather than the continuing struggle to remain competitive with price-reactive substitutes for paper products. Plastic products have displaced a significant portion of the product range of packaging and container firms, and the struggle in the primary product producing segments of the industry is to find new uses for pulp, paper, and paperboard. Many firms involved in the conversion of paper already have succumbed or have gone over to the "enemy" and have become producers of plastic products that are offered through traditional channels of distribution. The threat posed by plastics as a substitute for paper seems to be contained at the present time, and individuals in the paper industry appear confident that this challenge will continue to be offset with new applications for pulp and paper. One of the major successes of the paper industry in this respect has been the increased penetration of the packaging market with innovative applications of paper and paperboard to displace crates and pallets made of wood and glass and metals in the manufacture of food and other containers.

Opinion seems less sanguine in respect to the electronic media revolution, although opinion is that the effect of this revolution to date has been to increase the demand for business paper in its many varieties. No one expects this to be the final resolution and this is the principal threat expressed by individuals in the industry. Paper presently provides a sense of security to individuals raised under the influence of the Gutenberg revolution, but many believe this need will disappear as new generations are encultured by the electronic media. Some say that the increased demand for paper produced by computer application may last ten years and quickly diminish as modes of data storage are assured. The more vital punch card was an equally transitional phenomenon in the electronic computing age, with much the same crescendo effect of rapid rise in demand and then sudden collapse.

Much of the hopes of the industry bank on the intrinsic value of wood fiber as a resource with infinite applications. Wood fiber available for pulping has many of the characteristics of a by-product to the forest products industries, and thus it commands its application. Of course the same case can be made for many of the products that seek to substitute for paper. Nevertheless, our research indicates a positive view on the long-run ability of the industry to accommodate to the threat of substitutes. For example, in our survey three questions pertained to the importance of substitutes in restraining industry profitability. Only 7 percent believed that the threat of substitutes had influenced pricing to a degree that made it an extremely important influence on profits. Similarly, as to whether profits had been influenced in an important fashion by the actual introduction of substitutes for the products of the industry, only 9 percent believed it to have been an extremely important influence on profits. Finally, only 5 percent of the respondents believed that customers actually were substituting to a degree that had an important influence on industry profits.

Responses to each of these three questions appeared to be related to the segment of the industry in which the respondent was located, indicating that there was a differential influence of substitutes on the sectors. This finding is illustrated in the three exhibits, 5-9, 5-10, and 5-11, that depict the proportionate responses from selected segments of the industry. As can be seen, such sectors of the industry as the manufacturers of stationery and sanitary food containers appear to be more heavily influenced by the threat or actuality of substitution. The manufacturers of envelopes and corrugated cartons, on the other hand, expressed less apprehension about the threat of substitutes.

The issue of substitution was a topic of popular discussion among members of the various segments of the industry, and it was interesting to discover that a consensus exists. The leaders of the industry appear to have concluded that there will always be a place for paper and this opinion has been proclaimed and accepted. These proclamations paraphrase Senator Hammond's declaration of the compelling value of cotton by announcing that Wood Fiber is King. Of course the nature of pulp and cotton cannot be directly compared, but the analogy to such famous last words should not be overlooked. Raw materials and whole lines of business have been displaced in the past and undoubtedly will be in the future. This may not be the destiny of the paper industry, but an industry cannot survive by a complacency engendered by the antiquity of its products and process.

It was interesting that the responses to the survey did penetrate the consensus that existed in writings about the paper industry and the interviews that were conducted. Sectors of the paper industry do contain managers that feel a real threat from substitutes for their products, and apparently profitability in these sectors is at least partially under the influence of the threat or actuality of substitution.

Rivalry

In some respects the degree of rivalry that exists in an industry is a consequence of the threat of entry, the power of buyers and suppliers, and the threat of substitution; however, rivalry is a distinct feature of an industry. It is the way in which firms in the industry have separately chosen to deal with the environment and compete. The other four factors tend to shape the strategies that firms will employ, but enough options exist to allow divergent responses that permit industries that might otherwise be alike to have unique competitive structures. Industries have personalities that have been shaped by the shared competitive experience over time. Individual leaders have made their mark on the industry, and in turn managers in an industry are shaped by their experience in that industry. Industries in which there is little inward or outward flow of managerial talent will have characteristically strong personalities or cultures that would be difficult to change by simple maneuver. The consequence is a characteristic form of rivalry that exists among the firms that compete in the industry.

It should also be mentioned that the dynamic associated with rivalry includes the decision by some firms to seek out and exploit market environments that have less rivalry. Such segmentation can take a variety of forms, with the most familiar being a focus on the special needs of a particular group of customers, or the manufacture of a limited number of products or service in which a competitive advantage can be developed, or catering to a particular geographic market that other firms might have difficulty reaching. These are the terms of the focus strategies that we will discuss later, but these also are the historic basis for the origins of unique portions of an industry. It is also a basis for the specialization of an industry into elements that are no longer considered to be a part of a common competitive environment. In the terms of the population ecologists, two sets of firms are so unique that they can reside in the same environment without competing for the resources that are necessary for survival. Such firms may actually cooperate and enter into various forms of interrelationships without a sense of rivalry.

The intensity of rivalry is a function of the degree of concentration of firms in the industry and the balance among these firms in terms of size and resources. Rivalry is also influenced by the rate of industry growth because firms in slower growing industries will have to fight for market share to grow and prosper. Industries in which a higher proportion of costs are fixed will also tend to exude greater rivalry because of the need to fill capacity. One of the most significant factors producing rivalry is the lack of differentiation and the more commodity-like nature of the products that permit buyers to avoid switching costs. Industries in which increases in capacity must be accomplished in large doses because of the minimum efficient scale of operation will also have more rivalry because expansion will be chronically disruptive of the status quo. Industries in which the competitors are more diverse in background and culture also have the potential for greater rivalry because of differences in strategies that will be used; firms which are out of the same mold are more likely to think in the same way when it comes to competition. Finally, industries in which firms have high strategic stakes in the particular line of business and face high barriers to exit from the industry will have more intense rivalry.

If the above discussion seems familiar to those readers knowledgeable about the paper industry, it is because this industry is among those with this high intensity of rivalry. Prices are extremely volatile in the primary paper products sectors of the industry because of both the commodity nature of these products and the close relationship between profitability and the business cycle. The need for capacity operations drives these sectors of the industry and encourages every form of rivalry to gain outlets for capacity output. These sectors of the industry compete in national markets and increasingly in international markets. The latter has always been true in respect to market pulp, but it has become more prevalent in respect to paperboard, and few firms in the papermaking business lack some experience in international rivalry.

In most sectors of the industry firms are simply price takers, and the continuing lack of concentration of firms thwarts any attempts at price leadership. Almost every sector of the industry has firms that claim distinct advantages for their products and maintain that they hold competitive advantages of one form or another, but these are seldom translated into market advantages that permit distinct price schedules. As we shall see this also produced a great deal of ambiguity in the responses obtained with respect to competitive strategies. The Janus-faced character of the primary paper product sector obtained from our research may well be the result of a "hinge" at 93 percent capacity operation. Beyond this point the price to customers rises rapidly and gaining a source of supply becomes more and more critical to customers as full capacity for the paper industry is approached. Whatever marginal capacity exists is brought on line and quality assurance becomes a much less important factor to customers competing for a commodity in short supply. Below the "hinge" the door opens on a whole new set of strategies that focus on the differential and cost values of primary paper products in competition. No new customers appear as prices are reduced, as the use of paper is predicated on the nature of the economy; rather, customers look for values in the paper products they use to achieve their own competitive goals. Quality may be much more important than price, as the price of quality paper products becomes reasonable. Of course the "hinge" is made somewhat squeaky by customers who _must_ have a particular quality of paper products and reward paper firms with loyalty during downswings in the economy.

Four questions in the survey were directed at gaining information about the influence of factors associated with rivalry on profitability in the industry. The first of these asked whether competitive pressures had been increased in an important fashion by mergers and acquisition in the industry over the past five years. This addressed the issue of the effect of concentration on rivalry and profitability, and a clear majority of the responding business units felt that this was not a very important factor, with 52 percent indicating this to be unimportant or not very important and only 18 percent believing it to be an extremely or very important influence on profits. It appears from this that members of the industry either believe that concentration is not increasing or any increased concentration is not affecting profitability. However, when asked how important price cutting to gain market share had been in restraining profits in the industry, 62 percent believed this to be either extremely or very important, while only 16 percent of the respondents thought this was unimportant or not very important: one of the major manifestations of an industry with intense rivalry among competing firms.

The cost of gaining access to channels of distribution is often considered as an element of the competitive structure of an industry involved in rivalry, but when the survey asked respondents to indicate the importance of this factor on industry profitability, the average responses were lower than the other measures of interfirm rivalry: Some 59 percent of the respondents thought it was an unimportant or not very important factor, while only 14 percent thought it was either very or extremely important.

The final question asked how important to profits in the industry was the relative balance of strength among the firms in competition with the responding business unit. The majority (58 percent) supplied a response that indicated some importance.

Respondents in the various segments of the paper industry tended to answer the second and third of these questions in a different fashion, as shown in Exhibits 5-12 and 5-13. The manufacturers of envelopes and stationery are clearly distinguished by not having any respondents who ruled out the importance of price competition in their segments of the overall industry, while the papermakers were less likely to credit extreme importance to such types of competition. Neither the pulp or paper manufacturers or manufacturers of sanitary food containers credited any extreme importance to the influence of channel costs of profitability, whereas the folding box manufacturers were more likely to believe this to be an extremely important facet of rivalry.

Before concluding this analysis of rivalry, one major additional point deserves mention: the definition of rivals. In an industry relatively isolated from international events and trade, perceived and actual rivals are almost exclusively domestic firms. As the market and the industry become increasingly internationalized, new foreign rivals appear, although it may take many years before the domestic firms actually perceive foreign firms as major rivals. Furthermore, an internationally competitive industry is likely to perceive and to face significant foreign rivals than a less internationally competitive industry. The US computer industry is an example of the former; the US apparel industry is an example of the latter.

Viewed in the context of the world's pulp and paper industries, foreign rivals are not as significant to North American firms as they are to European firms. In fact, the former are major rivals of the latter, but not in reverse except in a few cases (e.g., the Scandinavians are rivals for US and Canadian exports to EC countries).

Furthermore, the characteristics of international rivals are more different than domestic rivals. They operate under different legal regulations, think and behave differently due to cultural differences (e.g., the Japanese versus the Italians), and have different cost and corporate structures, just to mention a few key differentiating factors. Therefore, as the pulp and paper industry becomes increasingly internationalized, new rivals and significantly different rivals will continue to emerge for pulp and paper firms in almost all countries. New and different rivals, in turn, should increase the rivalry within the industry and dictate changes in future corporate and industry strategies. The severity of the rivalry will depend on aggregate growth in demand and supply, changes in existing levels of the various forms of government protection, and the type and timing of new corporate strategies pursued. But one thing appears certain: Strategies successfully employed in the past are less likely to be successful in the future.

We conclude with Porter that the paper industry is best viewed as one with an intense competitive rivalry, but this conclusion is relative to segment of the industry. The paper industry has many component segments and the evidence collected indicates that the competitive structures of these components have some contrast that extends to the nature of rivalry itself. This segmentation is historic and there appears to be only minor transfer of experience among the segments that would permit the prediction of increasing similarity in a future period.

Summary of Issues in the Competitive Structure of the Industry

As previously indicated, the industry is best viewed as containing many parts. Some of these are quite distinct and best analyzed in isolation or in respect to other industries. Other segments of the industry are tightly connected with one another in a throughput process from the trees to the marketplace for paper products. The latter provide the greater analytical difficulty because these segments permit vertical integration, although this is not the choice of every firm. Some paper firms can only be understood in their character as fully integrated manufacturers of paper products. Other integrated firms must be evaluated in respect to the variety of forest products they are capable of manufacturing and selling. Although the previous section concluded that rivalry was intense in the paper industry, it does depend upon the sectors in which a firm participates and on the international competitiveness of a specific nation's industry or segments thereof. As a result it is difficult to make grand conclusions.

Furthermore, there does not seem to be one best way to compete and be profitable in the industry. Participation in the various sectors of the overall industry appears to have individual merit, and even within sectors different competitive styles have their adherents.

GENERIC COMPETITIVE STRATEGIES

It has been argued by Porter and others that there are three potentially successful generic strategic approaches to outperforming other firms in an industry, viewed as alternative designs for coping with the five competitive forces that face an industry. These three generic business strategies are overall cost leadership, differentiation, and focus. In the course of our study of the industry we looked for evidence of these three strategies and also included questions in our survey that asked about their use in the industry. Three questions in the survey asked the respondents to indicate the importance of each of the three strategies to their firm. Another question asked the nature of the geographic markets in which the firm competed, and a series of 30 other questions asked the respondents to indicate the importance of methods that are associated with business strategy. It was believed that this combined approach would provide insight on the competitive nature of the paper industry and reveal patterns that the forms of competition take in the industry.

The Statistical Addendum to this report provides detailed information on the results of the US industry survey regarding the methods and modes of competition. This portion of the chapter summarizes some of these findings and places them in the perspective of the larger study. As with the discussion of the competitive structure of the industry, there is some ambiguity in the results. In part this is the result of the difficulty translating theory into questions that tap the reality of the business world and the conflicting concerns of managers that often contradict the assumptions contained in such theories.

Overall Cost Leadership

Cost leadership strategy requires the creation of efficient-scale facilities, exploitation of the experience curve, tight cost control, avoiding marginal accounts, and having low levels of expenditures on R&D, service, sales force, and advertising. Firms using such a business strategy have the edge in the presence of strong competitive forces because they can often earn a profit in situations where others cannot. Firms using this strategy reduce rivalry to its simplest configuration and perform best under circumstances that match the terms of perfect competition.

The types of competitive advantage necessary for firms to exploit this strategy include a relatively high market share, favorable access to raw materials, products that are easy to manufacture, process skills, and a large product and customer base that permits volume production and sales. This strategy requires access to capital that permits continued updating of production facilities and expansion to absorb growth in demand. In some industries this is the only viable form of strategy that can be used because of the commodity nature of products, but in most industries it is possible to carve out niches where products can be differentiated or focused on certain markets.

While the paper industry cannot yet be characterized as one in which a pitched battle for market share exists, it does appear to lend itself well to the use of cost leadership strategies. This strategy tends to emphasize the importance of production methods in competitive situations, and the paper industry is process technology oriented. The industry also is capital intensive and requires high volume output to efficiently use the extensive plant capacity typical in the primary product sectors. A much-mentioned concern of firms in the paper industry in all countries is the ability to generate cash or otherwise obtain capital for the extensive refurbishing required to maintain competitiveness.

Firms that pursue a cost leadership strategy tend to be more backwardly integrated and attempt to gain control of as much of the process associated with the manufacture of their products as possible. This is a constant theme in the pulp and paper industry, particularly in the pulp segment, and it is believed that this drive for integration is to gain the control over costs that are vital to a cost leadership strategy. No sector of the paper industry is well known for elaborate channels of distribution and generally the costs of distribution are minimized.

Because pulp is generally considered a commodity product and one characterized by large production economies of scale, firms manufacturing this product are the most likely candidates for an overall low-cost leadership strategy. One of the more important competitive advantages firms hold in respect to this commodity is access to markets, and transportation costs and exchange rates are usually considered as the critical factors to success in these markets. However, as Chapter 3 indicated, the marketing of pulp involves other considerations including the relative advantage of sale versus internal consumption of the product for integrated paper firms. There are also many grades and types of pulp that lend themselves to different manufacturing processes, the production of which depends on many factors that are not commonly available to all pulp manufacturers. Certain types of pulp are critical inputs to particular paper product manufacturers, while only certain other types of pulp are fully interchangeable and amenable to substitution by pulp produced from recycled wastepaper.

The same types of features apply to paper and paperboard. Some of these products have commodity characteristics that would lend themselves to overall low-cost leadership strategies, while others are specific to certain uses by their nature and would permit other forms of strategy.

The noncommodity nature of paper, and to a somewhat lesser extent paperboard, is better appreciated than is the sometime noncommodity nature of pulp. This may be a matter of the volume of sales of the three primary products of the pulp and paper industry. More of the marketed pulp may be commodity in character so that an expectation of low-cost strategies exists, while manufacturers of paper and paperboard can be expected to have more choice of strategy through the products they might market.

The expectation was that other forms of strategy are more likely to be applied in the markets beyond the primary paper products. A low-cost strategy might be a feature of firms exploiting a limited market, but firms involved in the manufacture of converted paper products or containers were considered to be likely candidates for strategies other than low-cost, whether their markets are regional, national, or international. These were the expectations that led up to the interview and survey. However, the survey did not produce the expected results, even though certain tendencies were evident.

For the US industry overall cost leadership was a highly regarded strategy, with only a small group discounting its importance. Only 16 percent of the respondents stated that this strategy was unimportant or not very important to their business, while 44 percent indicated it was either extremely or very important. While this strategy was not held in quite as high regard as differentiation, it was considered much more important than a focus approach. However, there was no significant difference among firms based on the products they produced. The tendency was for firms in all segments of the pulp and paper business to have an equally high regard for the overall low-cost leadership strategy. The small differences among firms in the various lines of business are displayed in Exhibit 5-14.[1] Here we note that pulp manufacturers were most likely to indicate an extreme importance of cost leadership, but an equal number of units indicate it was only somewhat important. Undoubtedly the latter were firms that have sought to put greater emphasis on differentiation or focus strategies.

Outside North America, low-cost leadership strategies were also viewed as important but, in most cases, were also viewed as less feasible. As discussed in Chapter 3, virtually no other nation has the timber resource abundance and infrastructure to match the low fiber input cost of the North American industries and the resultant competitive advantage of the North American industries' primary products segments (which also extends through the downstream segments). As a result, they remain essentially price takers in the market and have to concentrate their strategies efforts either to minimize conversion and distribution costs or to follow differentiation or focus strategies. To be sure, firms in some countries invested abroad in timber-rich countries to try to gain access to lower fiber inputs (such as the Japanese firms) or made major efforts to develop alternative cost-competitive sources of raw material, such as recycled wastepaper (Europe and Japan). Yet such efforts, for the most part, were not sufficient to permit even these firms to become low-cost leaders in the international market.

Therefore, from a global perspective, all firms considered the advantages of adopting a low-cost strategy, but few outside North America could realistically exploit this strategy and, as a result, placed greater emphasis on alternative strategies.

Differentiation

A differentiation strategy involves the sale of a product that is perceived to be different from the products of competitors. Such differentiation can be in terms of product image, technological characteristics, customer service, form of product delivery, or any feature by which a product may be conceived by customers as different from that of competitors. This strategy seeks to insulate the firm from the five forces shaping the industry by distinguishing the firm from the industry and commanding a unique price in the marketplace. A firm employing this strategy also hopes to gain customer loyalty that will provide protection from firms hoping to enter the industry. Firms that successfully employ the strategy tend to view competitors as offering <u>substitute</u>, rather than competitive, products. In this sense a firm following a differentiation strategy is attempting to define a unique industry in which it is the single firm. If overall cost leadership strategies have roots in the economic theory of perfect competition, then differentiation strategies are a part of the neoclassical analysis of imperfect and monopolistic competition.

Firms using a differentiation strategy typically emphasize marketing skills in the same way that firms using a cost leadership strategy emphasize production skills. The emphasis in such firms is focused on product engineering and creative flair, rather than process engineering and cost control. These firms also have a stronger capability in basic research and have a reputation for quality and technological leadership. Firms employing a differentiation strategy will also often tend to be more integrated toward the market than toward raw materials, as these firms require greater cooperation from the channels of distribution.

As already indicated, it was expected that firms in the primary product sectors of the pulp and paper industry would have a strong penchant for cost leadership strategies while firms in the other sectors of the industry would lean more toward the use of differentiation strategies. It was thought that certain sectors such as wallpaper manufacturers and other specialty converters would be more especially involved in differentiation strategies and to a lesser extent the paper coaters, stationery manufacturers, and makers of envelopes. Yet, few paper firms beyond the manufacturers of sanitary paper products are well known for the types of marketing tactics that are called to mind by a differentiation strategy. Few firms deal in consumer and business paper products and those that do usually market such products through paper merchants. This tends to blunt the impact of differentiation strategies. Some paper firms such as Mead and Hammermill are known for more extensive forward integration that includes internal paper merchandising, but these are exceptions. The more typical customers of paper firms in all the realms of the industry are industrial customers who are knowledgeable about the paper products they use as raw materials and, as a result, often discount claims for differences and appreciate the opportunities to substitute products.

Given these features of the paper industry, the researchers were quite surprised by the results of the interviews and the survey. Respondents indicated that differentiation strategies were more likely to be of greater importance to their business than other strategies. Where 44 percent of the US respondents indicated that overall cost leadership was either very or extremely important to them, 85 percent indicated a similar response in regard to differentiation strategy, and only 3 percent thought differentiation to be unimportant or not very important (in contrast, only 16 percent claimed the same for overall cost leadership). Differentiation was the most popularly indicated strategy of the 361 business units responding to this portion of the survey. While it seemed that firms supplying sophisticated customers would have to place heavy emphasis on service and quality to compete, this did not always mean that the firms attempted to distinguish their products in these fashions. These conclusions are drawn from the information supplied by the responding firms in respect to competitive methods, which will be discussed later.

Less surprising than the US perceptions were those from non-US based firms. Given that they had little to no opportunities for much success in following a true cost leadership strategy, it was expected that they would emphasize more heavily the strategies of differentiation or focus. This expectation proved to be the reality. While the non-US based firm did seek to keep costs down, differentiation and focus were the cornerstones of their strategy, along with, in a few cases, the strategy of seeking greater government protection or financial assistance (e.g., France, Belgium, the Netherlands, and the UK in roughly that order).

Focus

The notion of focus is that firms' business strategies center attention on a particular buyer group, segment of the product line, or geographic market. A focus strategy is distinguished from both overall cost leadership and differentiation by emphasis on a particular target, rather than being a marketwide venture. The simplest example of focus is the way many firms serve local or regional markets with a geographic focus. Such firms can be visualized as specializing their attention on the needs of customers in particular localities. The same idea of specialization can be extended to particular products or the needs of particular customers in national and international markets, as well as regional markets. The essence of this strategy is the concept of carving out a niche in a larger market; the firm then seeks to concentrate its effort on the customers within this niche.

The concept of focus strategy is clear-cut until the question of how the firm will compete in this niche is addressed. An initial reaction is that a firm selecting this strategy has separated itself from competition by finding a protected niche in the environment, but in most cases there are still competitors. The regional distributor must compete with national and international distributors of the same product. The firm that specializes in a particular product must compete with manufacturers of broader ranges of products that include the particular product. The firm that focuses on a particular customer must compete with firms that include that customer as a candidate for their products, and the customer will view such competitors as potential suppliers. Therefore a focus strategy must have other features beyond those supplied by the competitive advantage of specialization, and firms that use a focus strategy typically use aspects of both cost leadership and differentiation within their niche or domain. In effect, a focus strategy acts as an added advantage to a firm employing either or both of the other strategies.

The paper industry is clearly rampant with firms using a focus strategy. Transportation costs for many products of the industry provide a competitive advantage to firms that specialize in particular geographic localities. The degree of geographic specialization that is practiced in the US industry is illustrated in Exhibit 5-15, which indicates the proportion of sales of the responding businesses in various types of domestic and international markets. Over 60 percent of sales are achieved by an emphasis on regional domestic markets, with national markets providing another 37 percent of sales. Many of the responding business firms indicated that all their sales were in regional markets and they made no attempt to compete internationally, let alone nationwide.

It is also clear that few firms in the paper industry set out to supply the complete paper needs of customers. The wide range of products offered by such firms as International Paper, Georgia-Pacific, Mead, James River, and Domtar is therefore unusual. Similarly, many firms that have characteristically broad lines of products, such as Crown Zellerbach, Great Northern Nekoosa, Hammermill, Mead, and Scott, themselves manufacture only small proportions of the products they market. It is much more characteristic for firms to specialize on products that result from a particular technological process. The heavy capital investment required in the manufacture of primary paper products dictates this focused approach.

It was less evident that pulp and paper firms concentrate on a select few customers, as the previous comments about buyer power indicated. In fact, firms in the industry seem to struggle to prevent their business from being dominated by a few customers. Thus there do not seem to be sufficient advantages to overcome the disadvantages of a customer-niche dependence. Firms in regional markets may be perceived as serving a limited number of customers, but it is clearly the regional advantage that is being emphasized in these cases, as these firms seek to develop new customers as regional conditions permit.

In spite of the strong case that can be made that focus strategies are an important part of the paper industry, the survey of US firms did not provide direct support for this position. The focus strategy was the least important of the three strategies in the minds of the US respondents as only 21 percent believed it to be a very or extremely important component of their business strategy compared to the 46 percent who said it was unimportant or not very important. This low level of importance accorded the focus strategy probably stems from the characteristics of the US industry and the low level of import competition. As noted previously, the US industry is in the best position to exploit the cost-leadership strategy, and it _does_ exploit it both domestically and internationally. Furthermore, the existence in the US of numerous very large integrated firms and large firms with broad product lines makes it more difficult to achieve success with a focus strategy: Market niches are more difficult to maintain.

However, outside the US (and, in general, outside North America), a different situation exists. A few countries have put in place significant import trade barriers, giving their domestic firms a protected market niche of sorts. On a somewhat wider geographic basis, the EC's absence of internal tariffs and quotas but maintenance of a system of common external tariffs and quotas provides EC producers with a variety of market niches which can be successfully exploited by a focus strategy or, on a wider scale, a differentiation strategy. The German paper manufacturers have followed these strategies with some success and, compared to most of their EC competitors, <u>considerable</u> success.

In a different manner, local cultural practices can afford a market niche and the opportunity for a successful focus strategy. A classic example is in Japan, where the domestic demand for the highly prized "washi" is serviced almost exclusively by the Japanese paper firms. Other examples exist, wherein, for example, cultural preferences for a certain quality of paper or paper with other features give an advantage to local producers who understand the preference and who have the capability of satisfying it.

COMPETITIVE METHODS

While the responses from company managements about the use of the three generic competitive strategies were generally not peculiar to the segments of the industry, responses to some of the questions about the importance of particular competitive methods to the strategies of firms were found to be significantly related. This section of the chapter first presents and analyzes these findings with respect to the US industry. (The complete results of these 30 questions are contained in the Statistical Addendum to this report.) Then, these findings are compared to the competitive methods of some of the firms in Japan, Canada, and Western Europe.

The United States Industry

The questionnaire survey of US firms asked them to evaluate the relative importance of 30 competitive methods, as enumerated in Exhibit 5-16. Each of these methods is generally associated with a particular generic strategy. For example, tight cost control and high efficiency are associated with a cost leadership strategy; quick response, high quality standards, and advertising and promotion are associated with the differentiation strategy; and narrow product line and service a specific, limited geographic market with a focus strategy.

If competitive theory is right, firms following a specific generic strategy should value highly and employ the competitive methods associated with that strategy and downplay the importance and not employ methods associated with the other generic strategies. However, the compelling logic of this argument/hypothesis was not borne out by the responses to the survey. On the contrary, a rather paradoxical picture emerged.

For example, the competitive methods consistently accorded greatest importance were all associated with a differentiation strategy: quick delivery and immediate response to customer orders, customer service, and high quality. However, three of the four methods accorded the <u>lowest</u> importance were also ones associated with the differentiation strategy: advertising and promotion, research, and ownership of patents or proprietary knowledge. Furthermore, nearly 80 percent of the respondents evaluated as very or extremely important tight control of selling, general, and administrative expenses--a competitive method generally associated with a cost-leadership strategy rather than a differentiation strategy! Thus firms in the US industry appear to use selectively the methods of differentiation while still attaching great importance to cost control--perhaps suggesting the inclusion in their minds of lower costs as a differentiating method of competitive strategy.

An alternative explanation examined was that these aggregate responses of the relative importance of competitive methods masked different relative importance evaluations by firms in specific <u>segments</u> of the industry. This possibly did, in fact, explain part of the paradox. For example:

1. Box manufacturers were more inclined to believe that designing or producing to order for individual customers were extremely important aspects of their business strategy, while none of the pulp manufacturers and few of the papermakers considered this method extremely important.

2. Manufacturers of sanitary food containers were much more inclined to highly evaluate the effective control of distribution channels (67 percent rated this factor as either very important or extremely important), while more than half of the firms in the envelopes and folding boxes segments assessed this factor as not very or not at all important.

3. The manufacturers of envelopes and boxes were more inclined to rate as extremely important serving limited geographic markets, whereas no responding manufacturers of paperboard and papercoaters shared this view, and 50 percent of US pulp producers evaluated this factor as not very important.

4. The pulp manufacturers were generally distinguishable by the importance assigned to competitive methods associated with overall cost leadership, while container manufacturers were similarly distinguishable by following methods associated with a focus strategy.

Yet despite these and other examples of industry segment patterns, there remained a number of anomalies and paradoxes difficult to explain, as illustrated by the responses from the papermakers. Their responses were so widely distributed that they defied categorization, suggesting that all three generic strategies were in use. That is, some employed the methods of overall cost leadership, others the methods of differentiation, and others the methods of focus. To a lesser extent, a similar observation could be drawn for several other segments of the industry.

What this situation suggests is that the US market is so large, the product offerings so vast, and the companies' competitive attributes sufficiently different that there is room for the pursuit of all three generic strategies within the overall industry and even within segments of the industry. Further support for this observation can be seen by looking at the "strategic profiles" of just the largest US firms (contained in Appendix 1 at the end of the report). Similar analysis of smaller firms would reveal even greater differences. Hence the complexity of trying to analyze competitive strategies and methods in any large US industry, not to mention a large global industry.

Canada

The Canadian industry's strategies were based largely on their accurate perception of a cost-price squeeze situation. Canadian production costs were rising faster than those in the US, while domestic and export prices were being held down by US firms. Because they could do little to influence prices upward, given sharper US competition, the Canadian industry sought ways to lower costs.

The main strategy employed by Canadian firms was increased capital expenditures to modernize existing capacity: making machines faster and more efficient, converting to more energy and fiber economical processes, and any other activities which would lower costs. Examples of the latter were pruning marginal operations (unprofitable businesses and inefficient mills and machines), salary cuts, and staff reductions. In a few cases, new facilities were brought on line despite existing industry overcapacity in certain sectors such as newsprint.

In addition to these domestic strategies, several Canadian firms acquired or expanded operations outside Canada. For example, Abitibi bought out Price, another Canadian company with hardboard manufacturing facilities in Mississippi, Ohio, North Carolina, and Indiana; established with Thomson Newspapers a joint venture newsprint operation in Georgia; and acquired a 50 percent interest in another newsprint facility in Louisiana. Consolidated-Bathurst bought Britain's Bowater Corporation's newsprint mill in Ellesmere Port in 1981 and put over $20 million into mill improvements. Particularly active abroad was MacMillan-Bloedel Ltd., with substantial timber ownership in Alabama and with lumber, plywood, wafer and particle board, newsprint groundwood, pulp, paper and paperboard, and corrugated plants in the US and UK in addition to Canada. It also owns part or all of several trading, wholesaling, and distribution firms in ten countries and recently acquired full ownership of Embrasca, a former joint venture with Brascan in Brazil. In short, MacMillan-Bloedel's strategy seemed to be to diversify and expand in every geographic direction.

Other strategies followed by some Canadian firms involved converting existing capacity to faster growth, more profitable grades of products. One example was Abitibi-Price converting newsprint capacity to uncoated groundwood, resulting in its becoming a leading producer of this product. Another example was Domtar's recent emphasis on fine paper and integrated containerboard facilities.

In sum, Canadian firms, with their historic competitive advantages accruing from vertical integration and cheap wood fiber supplies, pursued strategies designed to maintain their international competitiveness in the face of increased competition from US firms and decreased prospects for exports to the US and Western Europe. For additional information about Canada's leading firms and their activities, see the company profiles in Appendix 2 at the end of the report.

Japan

Unlike the North American industries with sizable supplies of wood fiber, fiber import dependency was an increasing problem for the Japanese paper industry, and this import dependency shaped many of the strategic moves within the Japanese industry. Especially in the 1970's, individual firms and groups of companies invested directly in timber-abundant countries to assure adequate sources of supply and to minimize historic price fluctuations of fiber supplied by unaffiliated foreign firms. Some ventures took the form of afforestation projects with test plantations designed to maximize fiber returns in the shortest possible time. Other operations were located in less developed regions with timber surpluses, such as Malaysia, Indonesia, Brazil, and Papua New Guinea. Others were centered in Oceania, Canada, and the US. Many of these latter were joint ventures with existing pulp and paper companies in the host country. See Exhibit 5-17 for more information about these overseas investments.

The three most active groups in the Japanese industry were (1) the Japanese Overseas Afforesting Association (JOAA) comprising 11 leading paper companies and one forestry company, involved mostly with overseas pulpwood planting; (2) MDI, a development company jointly established by Kohjin, Jujo Paper, Kanzaki Paper, Sanyo-Kokusaku Pulp, and Oji Paper, mostly operating in Southeast Asia for chip production; and (3) the Japan Brazil Paper and Pulp Resources Development Company, established by 18 pulp and paper companies including Oji Paper with C. Itoh and Co. to develop Brazilian timber resources and to produce chips and bleached sulfate pulp (750 metric tons per day).

These ventures were established as a hedge against rising imported fiber prices, the uncertainty of which continues to be one of the biggest disadvantages the Japanese industry faces, both on international markets and on its own domestic market.

The worst fears of the Japanese paper manufacturers were realized in late 1979 when Weyerhaeuser, a significant supplier of chips to Japan, suddenly raised its chip prices 67 percent over six months. This hike was quickly followed by other suppliers. Even though the prices had moderated significantly within three months (the following contract period), the increase in costs coupled with flat domestic demand and overcapacity caused an acceleration in the structural distress of the industry. Company profits dropped steeply and this decline did not show much improvement until late 1982.

The "chipshock" of 1979-80, as it became known in the Japanese industry, had several immediate as well as long-term effects. The Ministry of International Trade and Industry threatened to apply an unused trade regulation "forcing buyers to secure approval on their proposed import quantity, price, and port of loading in advance of formal import approval" to chip imports.[2]

Oji and Daio at this time were announcing plans to import from the US South, even though ocean transport through the Panama Canal is twice the distance from the US West Coast. "The Japanese firms chose a Georgia supplier . . . because they feared spot contracting with West Coast suppliers would help spur prices upward and create chip shortages in the region."[3]

While the immediate effect of "chipshock" was to encourage more intense interest in overseas investment, many of these investments began to look less promising as the domestic demand for paper dropped in 1980-81 and as economic recession began to slow GNP growth in Japan as well as in its principal markets. Also, early experience with plantation forestry in the tropics showed that the extremely high yields promised under laboratory-type controlled conditions were not forthcoming when the project was scaled up.[4] As a result, some of these overseas projects were abandoned or put on a "back burner" in 1982 and 1983.

Apart from these foreign investment strategies, the Japanese also followed the Canadian strategy of modernization and pruning both at the level of the firm and, as described in the previous chapter, at the industry level via the MITI-assisted industry rationalization plan. For example, Jujo Paper, the second largest pulp and paper company in Japan, has been spending (and will spend) over $40 billion on cost reductions and energy saving methods, improving quality, upgrading marketing, and consolidating its 38 subsidiaries.

Mergers, acquisitions, and consolidations were also key strategies undertaken by the Japanese industry. As a result of the consolidation of several of its businesses in which it held controlling interest, Daishawa Paper company became Japan's second largest "seaside, pulp to paper integrated company" by 1983. Similarly, in an effort to consolidate its top ranking in paperboard, Honshu Paper merged with Fukuoka Paper, Tonshin Paper, and Kanzaki Paper. The new company is now integrated from pulp to liner to corrugated box production. And, in an attempt to keep pace with

Honshu Paper, Jujo Paper merged with Senju Paper to make Jujo Japan's third largest paperboard company. (It is also now the second largest pulp and paper company in Japan.)

Even Japan's oldest and now largest paper company was not immune or insensitive to the merger/acquisition trend. As a result of several mergers, Oji is now integrated from commodity papers to high grade specialties. Its absorption of Nippon Pulp in 1979 typified its strategy: increasing and diversifying Oji's operations without adding any new capacity to the industry total.

The course of Japanese industry seems clear: to excel in relatively undeveloped export markets in East Asia for low-grade papers and to differentiate products in its domestic markets, yielding slowly to foreign competitors in fiber-intensive grades such as linerboard and high-strength newsprint. This strategy, coupled with structural pruning and reorganization, at this time appears to be the best--or only--course of action. Pressure to increase the price of the yen coming from political economic forces outside the industry will aggravate the already difficult position of the Japanese pulp and paper industry--an industry like sugar, cement, commodity chemicals, and textiles--where "prune," "shrink," "divest" signal a negative downward cycle in managerial talent, investments, and technology.[5] For more information about strategies of specific Japanese firms, see Appendix 3.

Footnotes

1. The differences depicted in Exhibit V-14 were not deemed to be significant. The components of the industry are more nearly alike than dissimilar in the indicated importance assigned to overall low-cost strategy. Even though the survey did not attempt to be a scientific sample of firms in the paper industry, the test of significance that was applied warns against belief that the responses of these firms indicate a difference in importance of a cost strategy. This same warning applies to the reported result of the survey in respect to the other two forms of business strategy.

2. "Woodchip import rises prompt Japanese action," World Wood, August 1980, p. 5.

3. "Japanese plan import of chips via Panama," World Wood, August 1980.

4. For example, Kawasaki Steel's Ipul-Ipul charcoal plantation in the Philippines produced actual yields of only about one third the promised yields based on early limited experiments.

5. Kenichi Ohmae, "The new technologies: Japan's strategic thrust," The McKinsey Quarterly, Winter 1984, p. 21.

EXHIBIT 5-1

Importance of New Domestic Competitors Threatening
To Lower Profits of Existing Firms

SECTOR	Not at All Important	Not Very Important	Somewhat Important	Very Important	Extremely Important
Pulp	16.7	33.3	0.0	50.0	0.0
Paper	20.9	32.6	20.9	23.3	2.3
Paperboard	45.5	31.8	13.6	9.1	0.0
Coating	14.3	39.3	28.6	10.7	7.1
Envelopes	14.3	25.0	14.3	32.1	14.3
Stationery	11.1	22.2	0.0	22.2	44.5
Corrugated Box	20.3	31.2	18.8	18.8	10.9
AVERAGE	20.4	30.8	13.7	23.7	11.3

EXHIBIT 5-2

Importance of Entry by Foreign Competitors
As a Restraint on Industry Profits

SECTOR	Not at All Important	Not Very Important	Somewhat Important	Very Important	Extremely Important
Pulp	16.7	33.3	16.7	33.3	0.0
Paper	23.3	48.8	9.3	18.6	0.0
Paperboard	65.3	13.0	13.0	8.7	0.0
Coating	34.5	34.5	20.6	6.9	3.5
Envelopes	82.1	10.7	3.6	3.6	0.0
Stationery	22.2	11.1	11.1	33.4	22.2
Corrugated Box	79.6	9.4	3.1	6.3	1.6
AVERAGE	46.2	23.0	11.1	15.8	3.9

EXHIBIT 5-3

Importance of Customers Having Other Sources of Supply
To Profitability in the Industry

SECTOR	Not at All Important	Not Very Important	Somewhat Important	Very Important	Extremely Important
Pulp	0.0	0.0	16.7	50.0	33.30
Paper	0.0	14.0	32.6	34.8	18.60
Paperboard	0.0	8.7	21.7	52.2	17.40
Coating	0.0	6.7	50.0	30.0	13.30
Envelopes	0.0	3.4	10.3	44.9	41.40
Folding Box	2.2	8.9	37.8	37.8	13.31
Corrugated Box	4.5	4.5	16.4	56.7	17.90
AVERAGE	1.0	6.6	26.5	43.8	22.26

EXHIBIT 5-4

Importance of Customers Threatening to Produce
Products of Your Industry

SECTOR	Not at All Important	Not Very Important	Somewhat Important	Very Important	Extremely Important
Pulp	16.7	50.0	33.3	0.0	0.0
Paper	58.1	34.9	4.7	2.3	0.0
Paperboard	56.6	26.1	13.0	4.3	0.0
Coating	60.0	13.3	20.0	6.7	0.0
Envelopes	44.8	44.8	3.5	6.9	0.0
Stationery	44.5	11.1	11.1	22.2	11.1
Corrugated Box	67.1	28.4	3.0	1.5	0.0
AVERAGE	49.7	29.8	12.7	6.3	1.6

EXHIBIT 5-5

Proportion of Raw Materials Supplied by the Three Largest Suppliers

Sector	Percent
Pulp mills	45.0
Paper mills	51.2
Paperboard manufacturers	33.4
Paper coaters	61.2
Envelope manufacturers	81.7
Sanitary paper products	41.3
Folding box making	79.2
Corrugated box making	71.1
Fiber can and drums	63.1
Paper merchants	60.0
Overall average	66.3

EXHIBIT 5-6

Importance of a Few Firms Controlling
Sources of Supply to the Industry

SECTOR	Not at All Important	Not Very Important	Somewhat Important	Very Important	Extremely Important
Pulp	0.0	50.0	50.0	0.0	0.0
Paper	23.3	44.1	18.6	14.0	0.0
Paperboard	47.9	43.5	4.3	0.0	4.3
Coating	16.7	23.3	26.6	16.7	16.7
Envelopes	22.2	25.9	18.5	7.5	25.9
Moldings	11.8	52.9	23.5	11.8	0.0
Corrugated Box	18.8	23.3	18.8	21.9	17.2
AVERAGE	20.1	37.6	22.9	10.3	9.2

EXHIBIT 5-7

Importance of Most Supplier Sales
Being Outside the Industry

SECTOR	Not at All Important	Not Very Important	Somewhat Important	Very Important	Extremely Important
Pulp	33.3	33.3	33.4	0.0	0.0
Paper	23.3	44.2	11.6	18.6	2.3
Paperboard	60.9	30.4	0.0	8.7	0.0
Coating	14.3	28.6	28.6	24.9	3.6
Envelopes	37.1	22.2	18.5	11.1	11.1
Moldings	35.3	41.2	17.6	5.9	0.0
Corrugated Box	43.0	23.1	15.4	12.3	6.2
AVERAGE	35.3	31.9	17.9	11.6	3.3

EXHIBIT 5-8

Importance of Suppliers as a Credible Threat to Produce Industry Products

SECTOR	Not at All Important	Not Very Important	Somewhat Important	Very Important	Extremely Important
Pulp	50.0	50.0	0.0	0.0	0.0
Paper	50.1	33.3	7.1	7.1	2.4
Paperboard	82.6	8.7	8.7	0.0	0.0
Coating	41.3	27.6	13.8	13.8	3.5
Envelopes	17.9	17.9	39.2	14.3	10.7
Fiber Cans	42.9	14.2	42.9	0.0	0.0
Corrugated Box	25.8	19.7	13.6	21.2	19.7
AVERAGE	44.4	24.5	17.9	8.1	5.2

EXHIBIT 5-9

Importance of the Threat of Substitutes On Profits

SECTOR	Not at All Important	Not Very Important	Somewhat Important	Very Important	Extremely Important
Pulp	0.0	0.0	50.0	50.0	0.0
Paper	19.0	35.8	28.6	9.5	7.1
Paperboard	17.4	17.4	39.2	21.7	4.3
Coating	6.7	26.7	49.9	10.0	6.7
Envelopes	44.5	29.6	22.2	0.0	3.7
Stationery	22.2	11.1	33.4	11.1	22.2
Folding Box	25.1	29.5	29.5	9.1	6.8
Corrugated Box	45.5	31.8	18.2	4.5	0.0
Food Cont.	0.0	10.0	60.0	0.0	30.0
AVERAGE	22.6	22.7	33.9	14.5	6.4

EXHIBIT 5-10

Importance of Substitutes Being Introduced
Displacing Products

SECTOR	Not at All Important	Not Very Important	Somewhat Important	Very Important	Extremely Important
Pulp	0.0	66.6	16.7	16.7	0.0
Paper	25.6	37.2	20.9	11.6	4.7
Paperboard	13.0	26.1	34.9	21.7	4.3
Coating	10.0	30.0	23.3	26.7	10.0
Envelopes	25.9	22.2	37.1	11.1	3.7
Moldings	0.0	35.3	35.3	11.8	17.6
Folding Box	22.8	29.5	29.5	9.1	9.1
Corrugated Box	35.8	40.3	16.4	6.0	1.5
Fiber Cans	0.0	28.6	0.0	57.1	14.3
AVERAGE	16.6	35.9	26.8	14.3	6.4

EXHIBIT 5-11

Importance of Customers Substituting
Industry Products

SECTOR	Not at All Important	Not Very Important	Somewhat Important	Very Important	Extremely Important
Pulp	0.0	0.0	50.0	50.0	0.0
Paper	19.0	35.8	28.6	9.5	7.1
Paperboard	17.4	17.4	39.2	21.7	4.3
Coating	6.7	26.7	49.9	10.0	6.7
Envelopes	44.5	29.6	22.2	0.0	3.7
Folding Box	22.2	11.1	33.4	11.1	22.2
Corrugated Box	25.1	29.5	29.5	9.1	6.8
Food Cont.	45.5	31.8	18.2	4.5	0.0
Fiber Cans	0.0	10.0	60.0	0.0	30.0
AVERAGE	22.6	22.7	33.9	14.5	6.4

EXHIBIT 5-12

Importance of Price Competition to Gain Market Share on Industry Profitability

SECTOR	Not at All Important	Not Very Important	Somewhat Important	Very Important	Extremely Important
Pulp	16.7	0.0	0.0	66.6	16.7
Paper	9.3	4.7	39.5	37.2	9.3
Paperboard	4.3	13.0	17.4	39.2	26.1
Coating	10.0	23.3	16.7	20.0	30.0
Envelopes	0.0	3.6	21.4	53.6	21.4
Stationery	0.0	22.2	22.2	22.2	33.4
Folding Box	6.6	13.3	31.2	20.0	28.9
Corrugated Box	2.9	2.9	19.1	38.3	36.8
AVERAGE	6.2	10.4	20.9	37.1	25.3

EXHIBIT 5-13

Importance of Cost of Channels of Distribution On Industry Profits

SECTOR	Not at All Important	Not Very Important	Somewhat Important	Very Important	Extremely Important
Pulp	16.7	33.3	33.3	16.7	0.0
Paper	22.0	43.9	24.4	2.4	7.3
Paperboard	36.4	45.4	18.2	0.0	0.0
Coating	27.7	31.0	31.0	6.9	3.4
Envelopes	25.9	37.1	22.2	11.1	3.7
Stationery	0.0	22.2	55.6	11.1	11.1
Folding Box	16.7	16.7	33.2	16.7	16.7
Corrugated Box	31.3	32.8	23.4	7.8	4.7
Food Cont.	30.0	30.0	20.0	20.0	0.0
AVERAGE	22.1	32.8	30.2	9.1	5.9

EXHIBIT 5-14

Comparative Importance of the Business Strategy of Overall Low-cost Leadership•

SECTOR	Not at All Important	Not Very Important	Somewhat Important	Very Important	Extremely Important
Pulp	0.0	0.0	50.0	0.0	50.00
Paper	0.0	20.5	45.5	29.5	4.50
Paperboard	0.0	17.4	34.8	39.1	8.70
Coating	3.3	13.3	40.0	40.0	3.30
Envelopes	0.0	6.9	48.3	34.5	10.30
Folding Box	6.5	21.7	37.0	23.9	10.91
Corrugated Box	1.5	14.5	36.2	34.8	13.00
AVERAGE	1.6	13.5	41.7	28.8	14.46

EXHIBIT 5-15

Proportion of Sales Reported for Various Market Sectors

	Pulp	Paper	Paperboard	Envelopes	Folding Box	Corrugated Box
Regional	48.8	32.0	53.2	77.0	78.3	85.9
National	37.2	61.1	41.5	23.0	18.7	13.0
Canadian	1.7	2.8	1.7	.2	.9	.3
Latin American	.8	1.8	1.3	.2	.1	.2
West European	10.2	1.1	.9	.1	.7	.3
Asian	1.0	.8	.6	.1	.1	.2
Other	.3	.5	.7	0.0	.2	0.0
Total	100.0	100.1	99.9	100.6	99.0	99.9

CHAPTER 6

SYNTHESIS, THE FUTURE, AND STRATEGIC IMPLICATIONS FOR THE UNITED STATES

Introduction

The previous chapters have analyzed several environmental trends that affect competition within and among national pulp and paper industries, the opinions of corporate executives concerning these and other trends, and the related strategies some of these companies have been or are pursuing. This final chapter first synthesizes previous chapters and evaluates the different strategic responses taken by the US, Canadian, Japanese, and European industries. The relative success of these strategies is examined in light of the social, political, and economic environments of each country and region, with some initial strategic implications drawn for the US industry. The remaining sections of the chapter look toward the future in terms of how expected changes in supply and demand, technology, government policies, and corporate strategies could affect the future international competitiveness of the US industry and what can be done to enhance it.

SYNTHESIS OF THE BASIS OF COMPETITION WITHIN AND AMONG NATIONAL PULP AND PAPER INDUSTRIES

The basic factors influencing competition within a domestic industry similarly influence competition in global industry: entry and exit barriers, buyer power, supplier power, availability of substitutes, and the nature of rivalry among firms in the industry. However, there are <u>additional</u> factors influencing global competition: major differences in availability and cost of factors of production, differences in the size of domestic markets, differences in societal values and capabilities, differences in government policies, and differences in the purchasing power of currencies.

These international factors must be considered when examining competition from a global standpoint because even though an intense rivalry might exist among domestic firms a protected national industry might not be competitive on an international basis. Stated another way, international competition in itself causes major pressures for adjustment on domestic industries except in the rare case of a country which is totally self sufficient (being the lowest cost producer in the world and producing all its needs) or one which is totally protected/isolated from international trade. Furthermore, the more internationally dependent a country's industry is, in terms of either imports or exports, the greater the impact will be on the industry of international developments.

International Pressures for Adjustment

The pulp and paper industries of the world have undergone considerable adjustment over the past four decades as a result of several major changes in their competitive environments. The rebuilding of the war-torn economies of Europe and Japan caused an increase in the demand for pulp and paper and brought with it major capacity expansions. The formation of the European Community (EC) in 1957 helped fuel the rapid economic growth in its member countries by dismantling trade barriers among its members, thus exposing each national industry to greater competition. The trade liberalization efforts among OECD members (the Dillon, Kennedy, and Tokyo rounds) had similar effects but on a larger scale in terms of countries. Ecological concerns led to environmental protection legislation in the US and similar concerns quickly spread to many other nations, generally altering capital expenditure plans and allocations, and raising production costs and product prices. The world energy shocks of the 1970's also had widespread but differential impacts in magnitude. Also, pulp and paper demand began to level off in the developed countries while surging in the more successful of the developing countries. Finally, differences in the magnitude and timing of national recessions and expansions had differential impacts on supply and demand conditions in each country and significantly affected world trade patterns and conditions, as did dramatic changes in exchange rates beginning with the breakdown of the Bretton Woods system in 1971.

Major Adjustment Strategies

As a result of changing competitive conditions domestically and internationally, the pulp and paper industries in virtually all countries experienced pressure for structural adjustment, changes in strategy, and often resorted to political action to slow or postpone the adjustment process. Adjustment strategies included seeking changes in trade protection and/or other forms of government assistance, minimizing production costs through modernization or international moves (foreign investments or international contracts), acquisitions and mergers, changes in degrees of vertical integration or diversification, and planned or unplanned industry restructuring. The choice and success of any of these adjustment strategies or combinations of them, however, were dictated and limited by national conditions such as domestic market size, stage of industrialization, and the international competitiveness of the national industry. These limitations typically determined whether the national adjustment strategies pursued were basically outward, internationally aggressive ones or inward, domestically protective ones.

COMPARATIVE NATIONAL ADJUSTMENT PRESSURES AND STRATEGIES

This section focuses primarily on the pulp and paper industries of the United States, Canada, Japan, and several West European nations. Their selection was due both to their importance in pulp and paper production, consumption, and international trade and to the different adjustment pressures they faced and how they reacted to these pressures. However, some mention is made of the pulp and paper industry conditions and adjustments in other regions and countries as appropriate. The discussion is organized by geographic region and the major countries within each region.

North America

Much if not most of what moves, shakes, and shapes world trends in the global pulp and paper industry is determined by North America, specifically Canada and the United States. North America is the largest consumer and producer of pulp and paper and, in volume terms, the largest exporter. In 1990, North America is forecast to have a 9 million ton pulp surplus and an 8 1/2 million ton paper surplus. As a region, North America has also experienced a 32 percent decline in world import share and an increase in world export share from 1966 to 1981. As a region, it has the greatest abundance of supplies of fiber, a large domestic demand, and the capital, technology, and management to exploit them.

Yet despite the historic and present pre-eminence of the region, the past two decades have had differential impact on the Canadian and US industries, and the future outlooks are quite different. The US industry has emerged in an even stronger position, while the Canadian industry faces a critical crossroads and is likely to experience a decline in relative pre-eminence.

The United States

While some US pulp and paper executives might not agree, from an international perspective most fundamental competitive factors favor the US industry. On a comparative basis, the United States has the most abundant supply of dense, accessible forests, a well developed infrastructure, the largest domestic demand for pulp and paper, relatively abundant and inexpensive energy and capital, increasing self-sufficiency, and little labor strife. Given these conditions, the US industry has been able to take advantage of huge economies of scale in the production of primary products (pulp, newsprint, etc.) and certain internal economies of scale via vertical integration. Yet even without vertical integration, the significant cost advantage of the US pulp and basic paper segments would still permeate through the downstream sectors of converted paper products.

To be sure, the US industry has experienced some government intervention "problems," notably in the areas of environmental controls, e.g., pollution and forest land withdrawals, and in antitrust, e.g., price fixing. Yet these problems were also experienced to some degree by pulp and paper industries in major competitor nations simultaneously or shortly thereafter.

Yet despite its favorable comparative advantage in virtually every sector of the industry, the US industry has remained essentially domestic in its strategy focus and influenced largely by domestic conditions and events. Only 10 percent of US domestic consumption is filled by imports, and less than 7 percent of US production is exported, even though the US trade surplus in this sector exceeds the aggregate production of all but a few countries. While there has not been any major change in industry strategy since the 1950's, there are some major shifts occurring now which have ominous implications for the pulp and paper industries of other nations. These strategic changes are resulting from the conjunc- tion of three factors: supply, demand, and oligopoly conditions.

On the supply side, the development of the southern US forest resource base has major implications for cost leadership strategy for the US industry. The US industry already has a dominant position in access to low cost wood fiber, and the development of the South should enhance this position even further. The most immediate impact will be import substitution of pulp and newsprint from southern sources for imports formerly coming from Eastern Canada. A second impact will be greater export competitiveness of virtually all US pulp and paper products for West European markets, again with Canada bearing the brunt of this potential as increased US exports take away Canadian exports to Europe. Finally, decreasing raw material costs should help the international competitiveness of the entire US pulp and paper industry--increasing the export potential and decreasing future imports.

On the demand side, the demand for pulp and paper is expected to increase faster <u>outside</u> the US than inside the US. Given the projected increase in US production capacity, the importance of capacity utilization for profits, and the smaller increase in domestic demand, much of the increased output will have to be targeted for export markets. The only major factors which limit this export strategy are a continued rise in the exchange value of the dollar, increased protectionism abroad, or lack of proper export planning by the US industry. These are all discussed in greater detail in the final chapter.

The oligopoly factor is likely to push the US industry to further internationalize its markets. There has been growing concentration in the US industry over the past 25 years, particularly in the primary products sectors. The larger firms have pursued most of their domestic options designed to gain a competitive edge over domestic rivals: vertical integration, product diversification, modernization of existing facilities, new mill construction, gaining access to lower cost sources of fiber, and some product differentiation in paper products and converted products. One major strategic alternative left is getting more aggressive internationally--getting higher prices and/or additional revenue from exports or simply using exports as a means of increasing capacity utilization and hence profitability, any and all of which could result in a competitive edge for the first few members of the oligopoly who pursue this strategy. Further, previous experiences of other oligopoly industries following this strategy suggest that once the initiators move more internationally, the other oligopoly members have little choice but to follow. If this pattern develops for the US pulp and paper industry, increased aggressiveness in international markets could prove highly disruptive for the pulp and paper industries in other countries, assuming protectionism abroad does not increase first.

Canada

The adjustment pressures affecting the Canadian industry have been more severe than for the US industry, and future strategic choices are more constrained. While Canada will continue to be the dominant surplus nation, its comparative advantage is decreasing, and it faces significant future difficulties in maintaining its export markets in commodity goods.

The crux of the Canadian pulp and paper industry's problem is its high export dependency. At high levels of capacity utilization, more than 50 percent of Canadian industry output is dependent on export markets, and the paper industry already exports 70 percent of her production. The Canadian industry is heavily oriented toward commodity products in general (especially pulp and newsprint), which by definition are highly price elastic. And as was previously mentioned, the development of the southern US resource base would in and of itself cause international competitiveness problems for the Canadians in terms of commodity exports to the US and Western Europe, even with the Canadian dollar being worth only 70 cents in the US. But such developments in the US are only part of the Canadian industry's problems; the others are internal.

Internally, the problems are many and diverse. Compared to its major competitor, the US, Canada has a less dense forest resource; slower crop rotation; more expensive transportation costs from forests to mills; older, less efficient mills; and a less developed infrastructure. In the past decade, labor problems have also increased significantly: The 1981 wage settlement for East Canadian workers which gave them 10-12 percent annual increases was twice the gains obtained by US workers, yet Canadian productivity is less than US productivity. Strikes by militant wood worker and paper worker unions worsened matters further in terms of production target shortfalls, lower capacity utilization, and hence, lower profits. Also during the past decade, Canadian government policies fluctuated concerning the industry. Through the mid-1970's, almost all major mill constructions received subsidies or other forms of assistance from either the federal or provincial governments (or both), and some provincial governments took over some troubled pulp and paper companies. In 1975, the government began buying back major public forests, thereby decreasing the supply and increasing the rent on available forest land. Environmental legislation in the later half of the 1970's caused a diversion of capital at percentages slightly higher than those in the US to comply with new standards. And, there continued to be disputes between provincial governments and the federal government over responsibilities for the forest products industry and the differential impact of acid rain on Canadian forests.

matched commensurately with increased regional production, causing a surge of imports. Asia's 14 percent share of world imports in 1961 increased 80 percent by 1981, and Asia currently accounts for over 21 percent of all import transactions in forest products. No other region experienced such a growth in imports.

There were several main reasons behind the import surge. While the region has plentiful forests, most are inaccessible. Also, the forest resources in much of Asia are primarily short-fiber, nonconiferous resources which to date have been less preferable for paper making. In addition, there are government and societal barriers present in most Asian nations concerning capital intensive pulp and paper mills. That is, there are preferences to emphasize more labor intensive industries and to use scarce and often expensive capital for more glamorous, high- tech industries.

Yet despite these generalized regional conditions, within Asia there exists a world class paper industry which is second in size only to the United States: the Japanese paper industry. Therefore, our discussion will focus on the adjustment pressures and strategies of this Asian economic giant.

Japan

The "economic miracle" of Japan brought with it a huge derived demand for forest products, and for pulp and paper specifically. In what has become typical Japanese industrial development style, the Japanese have developed a paper industry comparable in scale, efficiency, and quality to that in the US, and one that supplies virtually all of Japan's paper needs. However, although Japan is more forested than the US, its forests tend to be too young, inaccessible, and expensive to exploit. Pollution and high energy and land costs make the pup and paper industry an increasingly inappropriate industry for this increasingly high-tech country. As a result, Japan is highly dependent on foreign wood fiber--an import dependency that has shaped many of its paper industry's strategic moves. Therefore the adjustment pressures and strategies for fiber *import*

Internationally, like the US industry, the Canadian industry faced increased protection measures from the EC. Canada "voluntarily" constrained its kraftliner exports to the EC, was also investigated by the EC regarding alleged dumping of pulp, and later was involved in an EC investigation of an international newsprint cartel. To make matters worse, in 1984 the EC removed import quotas on Scandinavian newsprint, Canada's biggest competitor for newsprint exports to the EC.

About the only positive developments affecting the Canadian industry were (1) the weak Canadian dollar, which helped lessen the impact of rising costs, (2) the fact that its cost base, while rising, was still well below those in virtually all other countries, and (3) the growing demand for pulp and paper in the Pacific Rim countries of Asia, serviced substantially by West Canadian producers.

Given these mixed but preponderantly unfavorable pressures, Canadian strategies were more market maintenance than expansion oriented despite some increase in capacity. In fact, the driving force behind most Canadian strategies appeared to be to keep capacity utilization from falling further below already low levels, which for Canada meant aggressive exporting tactics. However, there seem to be fewer prospects for success in this strategy in the future, as two of Canada's major commodity export markets (the US and Western Europe) are likely to be eroded by US and Scandinavian producers. If so, this suggests a de-emphasis on commodity products in the future for Canada if her crucial export markets are to be maintained. However, Canada's competitive advantage in converted papers is currently less than that in commodity products.

Asia

In terms of demand for pulp and paper, Asia represents the fastest growing region. The increased demand for pulp and paper (and all forest products) has been driven by major industrialization programs and successes, particularly in East Asia, bringing with them increasing GNP's and per capita incomes. However, the increased demand has not been dependent Japan have been quite different from those for fiber and paper export dependent Canada.

As in North America, in order to achieve internationally competitive economies of scale in paper production, in Japan there has been a steady trend over the past 25 years of increasing industry concentration in several pulp and paper segments. Yet the cost structure of the Japanese industry remains quite different from North American industry due to less backward integration into domestic fiber production. Thus the Japanese industry must buy fiber on the expensive domestic market, import, or use recycled material--all of which are expensive, make fiber cost a higher percentage of production cost, and subject Japanese producers to price, exchange rate, and supply uncertainty.

As indicated above, the major adjustment pressure which has confronted the Japanese has been wood fiber shortages. The tremendously high cost of land in Japan, coupled with growing shortages of clean water, stringent pollution regulations, and high energy costs have made domestic supplies of fiber increasingly scarce and expensive. During the 1970's, many companies and groups of companies invested abroad in timber rich countries to assure fiber supplies for the domestic industry. Most of these ventures were also a hedge against rising imported fiber prices--one of the biggest uncertainties and disadvantages the Japanese industry faces.

Then came what was to become known in Japan as the "chipshock" of 1979-80, when major US chip suppliers to Japan increased prices by almost 70 percent over a six-month period. This increase in cost, coupled with flat domestic demand and overcapacity, accelerated the structural distress in the industry and led to formal government intervention and industry restructuring. These restructuring efforts, described earlier in Chapter 4, have occupied the center of Japanese industry thinking and strategy ever since.

The focus of the industry restructuring, carried out under the "administrative guidance of MITI, was to establish a "recession cartel" to work together to selectively reduce obsolete capacity (equally by company) and to move toward specialty, higher value added products and grades. The cartels of kraftliner, printing and coating stocks cut production by 50 percent of total capacity and pledged to scrap 80 percent of new capacity plans. New investments were limited to replacement or for increasing paper quality, process rationalization, or energy saving. The emergence of these cartels signaled the severity of the problems facing the Japanese industry, for much of the industry's history was characterized by intense competition, price cutting, and pre-emptive capacity additions.

The combined impact of import fiber dependency and overcapacity especially in the primary products sectors also led to corporate strategy changes regarding use of recycled material, forward integration, and diversification. Japan now uses a higher percentage of recycled waste paper than any other nation (40 to 50 percent vs. less than 20 percent in the US). Many paper companies have integrated forward into converted products and others have diversified into specialized papers or moved out of paper into related packaging and containers. Finally, the number of mergers and acquisitions within the industry increased as the restructuring occurred.

The Japanese industry's future strategic course is clear: Excel in the relatively underdeveloped export markets in East Asia for lower grade papers; differentiate its products in the domestic market; yield as slowly as possible to foreign competition in fiber intensive grades; and continue to structurally prune, reorganize, and modernize. While this strategic course is clear, its likely success is less clear.

Western Europe

As a geographic region, Western Europe is the largest importer of forest products, accounting for 50 percent of world imports--an amount more than double of any other region. And despite increases in production of both pulp and paper from 1960 to 1982 (87 percent and 125 percent respectively) and despite being a net exporter of paper and paper board,

Western Europe's imports (primarily of pulp) are so large that the region is a net importer of sizable magnitude.

Yet to discuss for very long the adjustment pressures and strategies of Western Europe's pulp and paper industry as a whole would be meaningless, for there are marked differences within the region. The Scandinavian countries are export dominant while the EC countries are import dependent. Thus our synthesis will be divided in terms of these two subregions.

Scandinavia

The pulp and paper industries of Sweden, Finland, and Norway influence most of what happens in Western Europe. Their low aggregate consumption coupled with strong export desires gives them a strong export dependency, in 1981 accounting for 80 percent of Western Europe pulp exports and 50 percent of Western Europe paper exports. In all three countries, the pulp and paper industry is an important economic sector and, as a result, receives substantial government attention. Their export aggressiveness has also received the attention of governments in the EC, but of a negative variety. In fact, the recent adjustment pressures on the Scandinavian industry have largely emanated from outside the region.

Given Scandinavia's ample timber resources and proximity to the EC, it has historically been a major supplier to the EC of pulp and paper. But when the pulp and paper industry on the continent hits hard times, the EC hits back at Scandinavia. In 1975, the EC froze import ceilings for 1976 at 1975 levels, not allowing the previously planned increase of 15 percent. In 1977, the Scandinavians were forced by the EC to "voluntarily" limit kraftliner exports to the EC. And in 1978, the EC again froze import ceilings on eight of 18 categories because of increasing unemployment and bankruptcies in the EC industries and because of alleged "unfair" Scandinavian government subsidies to their pulp and paper industries. Finally, in 1983 the EC raised tariffs on most grades of Swedish paper (4 percent to 17 percent) in response to what the EC viewed as an "aggressive" Swedish devaluation of the kroner by 16 percent in late 1982.

The "unfair" government subsidies argument was directed primarily at two Swedish government policies. Like most industries in Sweden, the Swedish pulp and paper industry was allowed to legally set aside part of their profits in a tax-exempt investment reserve. These funds could be taken out of the reserve for expansion purposes in designated times (e.g., recessions), effectively resulting in a portion of Swedish firms' profits being tax free. The Swedish industry did make use of this provision, putting money in through the mid 1970's, and taking it out in the later half of the 1970's. When these amounts were not sufficient, the Swedish government itself directly pumped in massive financial aid to save several companies. In addition, most Scandinavian governments provided low cost government loans to their industries to help mitigate the effect of rising energy costs in the 1970's.

Aggravating further EC concerns were the shifts in Swedish and Finnish industry strategies in the late 1970's. The Swedish industry's implementation of the "scissors" pricing strategy: raising export pulp prices more than the export prices of finished paper products in an attempt to increase paper exports and reduce pulp exports and capture more of the higher value added export product markets. Part of this strategy also resulted from a 1973 government initiated ban on all pulp mill expansions, coupled with directives aimed at better forest usage and location of future pulp and paper mills. The "scissors" pricing strategy was successful, as pulp exports declined 15 percent from 1976 to 1981 (they had increased 6 percent from 1971 to 1976) while paper exports increased 46 percent (compared to 17 percent from 1971 to 1976). The Finnish industry, meanwhile, was also moving to increase its exports of higher value added products in both pulp and paper. Pulp exports rose 63 percent from 1976 to 1981, while paper exports increased 35 percent. Yet despite an increase in exports of both pulp and paper, for Scandinavia as a whole there was a shift to paper, as paper exports in 1966 were roughly 80 percent of pulp export, but in 1981 were more than double the volume of pulp exports.

These combinations of government policies, shifting industry strategies, and the series of currency devaluations made the Scandinavian industry ever more export aggressive than they had been. More importantly, this put them in even greater competition with the struggling EC paper industry which most EC national governments were themselves trying to protect. About the only positive action foreign governments gave the Scandinavians was the EC's removal in 1984 of the tariff on Scandinavian newsprint.

The European Community

Like many other European industries, the 1970's was a difficult adjustment period for the pulp and paper industry, as the EC's net import position worsened. The EC's forest production base remained too small to meet the domestic demand for finished products, and the increase in domestic demand diverted production away from export markets. Thus imports grew while exports declined. With the notable exception of the Federal Republic of Germany, the EC nations increasingly sought to protect employment in their comparatively inefficient paper industries, but with little success. Employment in the EC industries declined by 34,000 from 1975 to 1980, and many long established companies were forced to close due to lack of international competitiveness, even in spite of increased EC protection.

In the United Kingdom (UK), the pressure for adjustment was particularly acute. UK pulp and paper imports are larger than domestic production: 90 percent of domestic demand for pulp is supplied by imports and 60 percent for paper. As imports surged by 51 percent in the 1970's, domestic output declined 23 percent and many mills closed despite increased use of recycled material and increased EC protection. The steadily declining exchange value of the pound sterling only aggravated matters further, making needed imports more costly with no countervailing favorable impact on exports (which were negligible in amount compared to imports). In addition, some firms invested abroad to avoid the high priced domestic pulp, and the persistent economic problems in the UK dampened future domestic expansion plans. Both worsened an already depressed situation domestically. In late 1975, the UK government set aside £25 million for capital investments over the next five years for the industry, but the amount was too little and too late for most of the industry.

If the situation in the UK was bad, the situation in France and Belgium was worse, and these two nations led the fight for increased EC protection during the past ten years. France also unilaterally increased the minimum prices for kraftliner sales and, after flirting briefly with an industry restructuring plan, finally instead developed a massive financial aid rescue plan. France partially subsidized loans totalling $73 million and $32 million to two large firms and loans of $65 million to three smaller firms. The Belgian government guaranteed the $50 million existing debts of the largest Belgian producer and $30 million of this firm's new debt. Regional governments injected even more funds into Belgian firms. Things were not much better in the Netherlands, where the government invested $1.5 million in an existing firm facing bankruptcy.

The only EC country which raised its voice consistently against government intervention domestically and worldwide was the Federal Republic of Germany. The German industry concentrated its strategy on structuring the industry around what it considered its major resource and competitive advantage: its skill in making paper and paperboard. It sought access to cheaper imported pulp (from Scandinavia and North and South America), while concentrating its exports on other EC countries where the absence of internal EC tariffs gave the Federal Republic of Germany greater export competitiveness over non-EC exporters. In addition, the EC export market was also more attainable than exports to non-EC countries. With domestic production exceeding imports by 60 percent and a more efficient, internationally competitive industry than any of its sister EC countries, the German industry fared better.

Other Regions and Countries of the World

Compared to North America, Western Europe, and in East Asia, Japan, few other regions or countries have been important in pulp and paper production, consumption, or international trade. Most suffer from key shortages in forests, infrastructure, and/or capital. However, there are two countries which still showed significant potential: the USSR and Brazil.

Until fairly recently, Brazil's vast tracts of tropical forests were largely inaccessible. This situation and the lack of domestic capital to finance capacity expansion resulted in less than internationally competitive scale economies and only marginal exports. However, in the 1970's capital influx surged from foreign bank loans and foreign direct investments, causing a subsequent surge in production capacity. Yet rapid internal economic growth consumed most of the expanded output until the 1980's. Foreign debt repayment burdens caused recessionary pressures in Brazil and also increased the need for exports to generate foreign currency. The expanded pulp and paper production capacity was accordingly shifted away from domestic sales and toward export markets in an aggressive manner. With additional improvements in infrastructure and forest management sophistication, Brazil's vast timber resources are likely to provide a major source of world pulp and paper in the future.

An even greater potential exists in the Soviet Union, which contains roughly 60 percent of the world's coniferous forests but produces only 6 percent of world volume of forest products. As discussed in Chapter 3, the Soviets have embarked on an ambitious program to boost processing capacity. They plan by 1985 to boost pulp capacity by 35 percent, paper capacity by 25 percent, and board capacity by 44 percent--all in an effort to avoid being a net importer of pulp and paper by 1990, a humiliating prospect for the world's most timber rich country. However, profound weaknesses in supporting infrastructure and other, more pressing industrial goals are likely to hinder the successful attainment of the development goals for the pulp and paper industry. In addition, most capacity expansion will be devoted to rising domestic consumption, rather than export markets. Thus the Soviet industry poses little threat to the export markets of the North American industry despite its geographic proximity to Western Europe. Of the other Eastern European countries, Poland and Czechoslovakia also have potential for becoming major suppliers, but they too face significant development problems.

Comparing the major pulp and paper countries discussed above, they can be categorized into four groups based on changes in their domestic capacity (increasing or decreasing) and changes in the future net trade position (improving or worsening), as depicted in Exhibit 6-1. Group 1 includes the US, Scandinavia, and Brazil: increasing capacity and improving net trade position. Group 2 includes Canada (especially eastern Canada), the USSR, and Japan: smaller increases in capacity but worsening net trade. Group 3 is comprised of the Federal Republic of Germany: pruning capacity but an improving trade position (in paper). Group 4 includes the UK, France, Belgium, and the Netherlands: declining capacity and worsening trade position.

The problems facing Group 4 are accentuated by the actions of Group 1 and, to a lesser extent, Group 3. As for Group 2, Canada's problems are also impacted by the actions of Group 1 (displacing Canada's historic export markets) while those of Japan and the USSR are affected more by internal conditions. The future success of the Federal Republic of Germany (Group 3) is largely dependent on its tariff and quota free access to EC markets and its specialization in finished papers.

As for the US industry specifically, it has become even better positioned in terms of international competitiveness. A reduction in imports is virtually certain, and a major increase in exports is certainly possible unless foreign countries increase trade protection or provide other forms of intervention/interference in the market. It is to these future developments and related strategies we now turn in the concluding sections of this chapter.

THE FUTURE

While the past and current international competitiveness of the US pulp and paper industry are of some interest, of greater interest to corporate managers and government policy makers is <u>future</u> international competitiveness. Therefore, in this final section of the study, we look toward the future in terms of expected changes in supply and demand, technology, and government policies; how these changes could affect the future international competitiveness of the US pulp and paper industry; and what could be done to increase the international competitiveness of the US industry.

To be sure, forecasting the future is an imprecise, difficult, and hazardous task. It involves estimating changes in the direction, magnitude, and rates of a host of interrelated developments, any errors in which can alter significantly the expected results. Nonetheless, if business and government are to be proactive rather than reactive, a future orientation based on reasonable forecasts is essential.

In looking toward the future, we assume that there will be no new factors emerging which will have major impact on the international competitiveness of the pulp and paper industry. By this we mean that those factors which have shaped past competitiveness will continue to do so in the future even though each factor's relative importance may change.

Forecasts of Demand and Supply: The US vs. the Rest of the World

Demand

The greatest future growth in demand for pulp and paper products is expected to take place outside the US, and particularly in the newly industrializing countries. In the past, demand for pulp and paper has closely followed economic growth, industrialization, and per-capita income, and we see no change in this historic relationship. If it holds, then the slowdown in industrialized nations' growth rates and the acceleration of growth rates in LDC's will begin to alter the historic demand patterns for pulp and paper, shifting gradually toward LDC's.

However, while demand will probably rise faster in LDC's, the absolute volume demanded will remain larger in the industrialized nations, and in the US particularly, for many decades. In addition, the growth in demand for certain paper products (particularly those in the early stages of their product life cycle) will be higher in the US than elsewhere. Thus the US market should continue to dominate the world market for the rest of the century. This forecast should not suggest that demand growth and market opportunities outside the US should be considered unimportant or insignificant. Quite the contrary; they are and will become even more so in the future. It should suggest that foreign markets have great potential which should be exploited, but not with the same degree of effort and importance as the domestic market--at least in the near future. In addition, as subsequent sections point out, greater exploitation of export opportunities will be increasingly important for US firms in terms of their competitiveness in the domestic market as well as in the international market.

Supply

On the supply side, the largest single country increase in capacity is expected to occur in the US, prompted by the Southern expansion and the modernization of production facilities throughout the US. The potential increase in US capacity may in fact be larger than the total increase in production capacity in the rest of the world. Other capacity expansions should occur in Canada, Scandinavia, and Japan, but largely from modernization of existing facilities rather than new facilities. Some new capacity will likely be added in developing countries, but will be small in magnitude. The only two potentially large impact countries whose future capacity expansion possibilities remain uncertain are the USSR and Brazil. Brazil will need to attract outside capital, which is problematic at best, while the USSR is likely to allocate its capital in more vital sectors.

One thing looks certain however. Based on recent and ongoing FAO studies, there will not be any world shortage of timber in the foreseeable future--the key raw material input for the pulp and paper industry. In the short term, the major problem will most likely be to carefully manage new capacity with volatile demand. In the long term, the supply factor will likely become more important, particularly in terms of access to timber. This long-term problem can be lessened, however, through continued improvements in forestry methods, recycling, and production efficiency.

Industry Segments

Unless a firm is vertically integrated and produces products in every segment of the pulp and paper industry, what happens to aggregate supply and demand for pulp and paper is less important than what happens to supply and demand conditions for the specific product it makes. Therefore, somewhat more disaggregated forecasts are needed.

Due to the expected increase in global demand for all paper products, the derived demand for pulp is expected to increase in virtually all countries. However, only a few have the natural resource base (timber) sufficient to have any major impact in terms of pulp supply: the US, Canada, Scandinavia, the USSR, and a few developing nations. And, as was pointed out in the previous section on aggregate supply, only the US, Canada, and Scandinavia are likely to increase pulp supply significantly enough to impact aggregate supplies.

A similar scenario seems reasonable for newsprint: rising demand globally, but major increases in supply coming primarily from the US, Canada, and Scandinavia. Other developed nations expected increase in newsprint supply will depend largely on increased imports of pulp and from increased use of recycling. In developing nations, some will be able to expand their exploitation of national timber supply (e.g., the Philippines, Indonesia, Malaysia, and Brazil) but most will have to import pulp if they want to increase local production of newsprint.

For most finished paper products, the absolute volume demanded will remain largest in developed nations, with the US continuing to dominate, despite higher expected growth rates in developing nations. On the supply side, Europe and Japan are expected to increase significantly their supplies as they focus their future production efforts on finished paper products and away from primary paper products. Canada should also be expected to move in this direction, albeit more slowly because of less urgency.

For specialty papers, the largest increases in demand and supply are expected to be in the US, which should also retain its number one positions in volume as well. Japan and the Federal Republic of Germany should follow in second and third positions respectively, in terms of both supply and demand. Comparatively little increase in demand or supply of specialty papers is expected in developing nations.

Forecasts of Changes in Technology and Government Policies

Major technological developments can significantly alter the supply of, and demand for, pulp and paper products. Technological improvements which lower production cost could lower prices and improve competitiveness, or result in higher profits, increasing the stability of the industry. On the other hand, technological developments which improve the substitutability of other products can have the reverse effects. In looking realistically toward the future, however, no major technological breakthroughs in either area are expected. Rather, incremental changes in existing trends are expected to continue:

1. Improved forestation techniques to lower cost of raw material input.
2. More efficient recycling techniques.
3. Lower production costs for pulp and paper.
4. Some continued substitution of paper products for other products, e.g., textile fabrics.
5. Some continued substitution of other products (e.g., plastics) for paper products.

Perhaps relatively more important changes can be expected in government policies, particularly those outside the US. The most likely and significant is an increase in protectionist measures by the European Community. Continued declines in the international competitiveness of the EC industry should cause continued reductions in employment, profitability, and the number of firms, which, in turn, will likely lead to an increase in EC protectionist policies, especially for newsprint and finished paper products. While Germany can be expected to argue against such policies (based on its historic commitment to free trade), other EC members' wishes will likely prevail. Furthermore, the German industry's comparatively strong international competitiveness within the EC would result in a favorable environment for German expansion within the EC market--currently its major export market--perhaps tempering Germany's position against increased EC protectionism.

Should the expected increase in EC protectionism occur, Canada and the Scandinavian countries will likely retaliate with increased government intervention to support their pulp and paper industries which rely heavily on exports to the EC. Even without increased EC protection, the expected increase in the US industry's international competitiveness could well prompt the Canadian, Scandinavian, and Japanese governments to intervene with additional supportive measures for their respective pulp and paper industries. However, current trade frictions between the US and Japan would likely mitigate any major increase in Japanese intervention.

As for the rest of the world, import substitution policies in developing nations are likely to continue and expand, especially for products beyond the primary sector. Such increases in protection, might be less severe than those by the EC due to smaller volumes of imports, less viable opportunities for import substitution, and less government emphasis on the pulp and paper industry compared to other industries. However, because the demand growth prospects are so high in developing nations, especially in the newly industrialized countries, increased protectionism of developing countries' pulp and paper industries may prove more significant in the future.

STRATEGIC IMPLICATIONS FOR THE US PULP AND PAPER INDUSTRY

For decades, the US pulp and paper industry has dominated global production and consumption. Based on the information gathered for this study, there is no reason to suggest that the historic dominance will wane in the foreseeable future. In fact, US dominance will probably increase in the future. Current and near-term future efforts to modernize existing facilities and further develop the Southern US resource base will further increase the US industry's competitive advantage at home and abroad. No other country has the natural resource base and available capital and managerial skill to keep pace with the US industry.

Yet despite this comparatively enviable position, there remain some serious potential problems. First, it is highly doubtful that future US demand can absorb planned future increases in US production capacity. As a result, export markets will become increasingly important for US firms in order to keep capacity utilization at sufficiently high levels. As has been pointed out, the future US export potential based on comparative economic advantage may well be hindered by increased import protectionism abroad,by increased government subsidization of foreign industries who also compete for the same foreign markets, and a continued strong dollar. Furthermore, the potential loss of US market share facing the Canadian industry may similarly lead to Canadian government intervention in its industry, mitigating the potential import substitution effect for the US industry and hence also causing capacity utilization problems in the US. Should the projected increase in US capacity occur and foreign government intervention policies increase, the US industry could face serious capacity utilization problems and with them, decreased profitability, employment, and number of firms. While most of this problematic scenario will initially affect the larger US firms, it would subsequently affect smaller US firms as the larger firms seek to keep capacity utilization high by aggressively seeking more US business and lowering prices. Thus smaller US firms must seek market niches to enhance their prospects for survival or growth.

For larger firms, a continuation of existing strategies is advisable and warranted, with greatest attention needing to be focused on other large US and Canadian competitors' moves. This is especially true for capacity expansions because the biggest imminent threat/problem is adding too much capacity at once. Other critical strategies involve the industry's relation with the US government:

1. Obtaining government pressure to reduce existing import barriers abroad and/or keep them from rising further.
2. Monitoring "unfair" trade practices of competitor nations' industries in terms of their export markets to the US and to

countries where the US industry is also exporting.

3. Keeping US capital, energy, and regulatory compliance costs at a competitive level with those of other major countries.

4. Keeping the foreign exchange value of the dollar from rising further, which, if unchecked, would seriously impact US export potential.

Points 1 and 2 are related to an expected increase in import protection by the EC, possibly Japan, and some developing nations. Increased tariffs, tighter quotas, and larger subsidization schemes are quite possible, putting strong competitive pressures on major producing nations which export heavily to these countries, such as Canada and Scandinavia. Should the expected increase in protectionism occur, Canadian and Scandinavian producers may have little more recourse than to focus increasingly their attention on less protected markets such as the US, and themselves become more protectionist of their own domestic industries. Should their industries become severely impacted, one can expect them to become very aggressive marketers in the US, perhaps aided and abetted by new government intervention schemes in their countries--at considerable potential peril to the US industry.

Points 3 and 4 are important to the US industry with or without an increase in protectionism abroad. Regulatory compliance costs have become fairly equalized among the major producing nations. Any unilateral increase in US regulatory compliance costs, therefore, could damage the international competitiveness of the US industry both at home and abroad (via exports). Similarly, the currently over-valued dollar has already adversely affected US exports. Its further rise would clearly aggravate the US industry's international competitiveness, while a future decline would have the reverse impact.

Given the existing international competitive advantage of the US industry, and that much of the future expansion of US industry capacity will have to be export market focused, maintaining and enhancing the US industry's well-being will require more US government support internationally in the areas enumerated above.

Finally, all firms in the US pulp and paper industry must put much greater emphasis on marketing than they have done historically. If there is anything the past two decades of increased international competition have shown, it is that success in the market depends increasingly in successful marketing. Neither product superiority nor lower production costs are any longer sufficient, and the pulp and paper industry is no exception. Nor does it make any difference whether a firm is selling domestically or internationally. Improved marketing research is clearly necessary: identifying new products, new buyers, new market locations, new product uses, and new patterns of consumption and buyer preferences. But even improved marketing research, while necessary, is not sufficient. Marketing research doesn't sell products. Salespeople sell products. If a salesforce doesn't have the appropriate degrees of commitment, knowledge, expertise, and top management's support (financially and otherwise), the sales result will be disappointing at best, and potentially ruinous at worst. While this observation pertains to both domestic and international sales, it is especially crucial for the latter. Given the geographic, cultural, and political advantages possessed by foreign firms in their own domestic markets, US pulp and paper firms must minimally quadruple their current marketing efforts and activities in both quantity and quality. Foreign markets and customers are more difficult to understand and to penetrate and, as a result, demand higher, more sophisticated marketing efforts. And a substantive part of what they demand is face-to-face interaction and corporate commitment at levels well above those they have previously experienced from US firms.

CONCLUSIONS AND CAVEATS

The US pulp and paper industry is not without its problems. Competition is intense, demand is cyclical, many manufacturing facilities are outdated or not well located, and industry profitability is low. Yet despite these problems, compared to the pulp and paper industries in other countries, the US industry is in relatively good shape. The fact that the United States is the largest consumer and producer of pulp and paper products gives the US industry a largely self-sufficient position. The immense and accessible forests in the United States provide a substantial natural resource base, and US forestry practices are among the most efficient in the world. There is also an ample supply of capital and managerial talent on which to draw, even though US capital costs may be higher than capital costs in other countries. Most production facilities are technologically advanced and many achieve substantial production economies of scale. Furthermore, regulatory policies and compliance costs are not out of line with those of most major competitor nations. In short, the US industry's international competitive position should strengthen from an already existing dominant base.

Potentially hindering this forecast of increased international competitiveness of the US pulp and paper industry are international factors largely outside the direct control of the US industry: likely increase in trade protection and other forms of government intervention abroad, increased aggressiveness by Canadian and Scandinavian producers, and the value of the dollar. All of these international factors would be less potentially damaging to the US industry if it were not planning to increase significantly its capacity, a significant percentage of which will have to be export market oriented. If substantial US capacity additions come on stream when domestic demand slackens and export markets are less accessible, the US industry could experience serious financial problems due to excess capacity.

In sum, the US industry will have a very solid base, in comparative economic terms, on which to increase its future international competitiveness. Unless the comparatively weaker industries abroad are so damaged by increased US competitiveness that they succeed in obtaining their governments' intervention in one form or another, the outlook for the US industry is favorable. Thus the US industry must exercise caution and possibly restraint in adding too much new capacity too quickly, too simultaneously.

Finally, the US industry must significantly increase and improve its marketing ability and activities, both at home and abroad. Failure to do so will erode the economic advantages it has over foreign competitors and will become the proverbial Achilles' heel that often causes even the most physically competitive runner to lose the race.

EXHIBIT 6-1

COMPARATIVE FUTURE POSITIONS

FUTURE NET TRADE POSITION	CHANGE IN DOMESTIC CAPACITY: INCREASING	CHANGE IN DOMESTIC CAPACITY: DECREASING
IMPROVING	United States Scandinavia Brazil	West Germany
WORSENING	Canada USSR Japan	UK France Belgium Netherlands

APPENDICES

NOTE: Background material about the largest US, Canadian, and Japanese pulp and paper companies was the first information collected for this research study and served to initially identify major differences in corporate strategies, structures, performance, and emphases. Summaries of this information are contained in these appendices.

In reading the appendices, however, please bear in mind that the information presented applies only up through 1983 (and in some cases, 1982) -- the most recent data available when the research study began. Therefore, the appendices contain information which is now considerably dated. Nonetheless, the information should still be enlightening and useful for a comparative, albeit dated, perspective.

APPENDIX 1

UNITED STATES COMPANY PROFILES

Boise Cascade

Boise Cascade is a highly integrated company with about one third each of its sales in pulp and paper, containers, and wood products. It went through some corporate restructuring in the late 1970's, before the the ensuing recession, and as a result went into the recession much leaner than many of its competitors. Boise had 10.8 percent of the uncoated papers capacity in the US, a lucrative line which can be expected to improve as business improves, with resultant demand for business and computer papers. The uncoated market is fairly concentrated as well, so Boise's relatively large market share carries correspondingly large market power.

Boise is also a leader in American newsprint, with 4.8 percent of US capacity. This is a product which is relatively resistant to business downturn, with demand quite inelastic, but newsprint capacity in North America and the world is thought to be considerably greater than demand. The possibility for downside pressure over the near term on newsprint prices, and profitability, is high.

In 1983 Boise had 9.3 percent of the concentrated (94 percent in the top ten producers; Boise is number six) coated groundwood publication papers. This is another relatively recession-resistant market but one in which price competition makes it worthwhile to be the low-cost producer. Rising output of newspaper supplements and direct-mail catalogs makes coated groundwood a probable high-growth market, which Boise is well-positioned to take advantage of.

Boise is 46 percent fiber sufficient and owns or controls 7.6 million acres of timberland in the US and Canada. In 1982 Boise was the ninth largest firm in the world in gross forest product sales and was eighth in the US, with sales of 2.9 billion. Seventy percent of its revenues in 1982 were generated from pulp, paper, and converting. This is a shift away from its more normal product mix of 62-64 percent, indicating Boise's flexibility to shift production away from recession-sensitive wood products as needed.

In 1983 Boise had joint ventures in the US, West Germany, and Austria. Its labor costs overall were about 24 percent of sales, slightly below industry averages, and its average plant age is estimated at seven years. At the end of FY 1983, debt to capital ratio was 31 percent, including $75 million in capitalized leases.

Champion International

Champion International is a major manufacturer and distributor of building products, paper products, and packaging. It is well integrated, with three million acres of timberland to provide 45 percent of its fiber needs for various paper mills, wood-processing plants and sawmills, and converting plants. Recent capital expenditures have trimmed unit operating costs and increased vertical integration. In 1982 Champion was the fourth largest paper manufacturing firm in the US and in the world, ranked by gross sales, which totalled $3.7 billion. Sixty percent of its sales were in pulp, paper, and converting, with the balance in wood products.

In 1983 Champion was the owner of 6 percent of the US bleached paperboard capacity. For this highly concentrated segment (the top ten producers have 91 percent of total capacity in the US), sales are expected to continue strong, since bleached paperboard is heavily utilized for food packaging. There is some competitive pressure from recycled paperboard, which may slightly erode concentration market power.

Champion has been the second largest American producer of uncoated freesheet printing and writing papers. This is a strong growth market, given the continuing rise in office copier use. Uncoated stock is also used for computer applications and other expanding office uses, with strong expansion expected as the economy improves.

Champion's linerboard capacity ranked fifth in the US. Linerboard has been depressed in the past due to world overcapacity and lowered industrial activity in the 1980-82 recession. As industrial demand for packaging and boxes improves, so will Champion's profitability in this line. Champion also ranked sixth in the US in corrugating medium capacity, with most of its liner and corrugating medium going to its own box plants. About 19 percent of Champion's sales were in packaging, including boxes and corrugated materials.

In 1983 Champion had foreign operations in ten countries, with manufacturing facilities in three. Its labor costs were 22 percent of sales, well below industry average, with an estimated plant age of seven years. At the end of FY 1983, debt to capital ratio was 35 percent, including $348 million for capitalized leases.

Crown Zellerbach

Crown Zellerbach is a fully integrated forest products firm with 88 percent of its sales revenues generated by its pulp and paper segments through manufacture, converting, and distribution. Crown Zellerbach manufactures paper packaging, office products, flexible packaging, and tissue, as well as market pulp and solid wood products. Distribution is an important business for Crown, and a new venture is currently being launched in computer suplies called EC-ZEL, which was projected to have distribution centers in major cities in 1984. The firm is integrated backward as well, with 3.3 million acres of timberland owned or controlled in Louisiana, Mississippi, Oregon, Washington, and British Columbia. These timberlands provided about 50 percent of the company's total fiber needs.

Sales in 1983 were $2,709 million, the lowest for the company since 1978. From 1980 to 1983, significant corporate restructuring took place including the replacement of the CEO, divestiture of marginal and outlier businesses, cost cutting, and staff reductions, along with implementation of a major capital investment program in both wood products and fiber businesses. The new Crown Zellerbach, with updated facilities, improved product line, and a lower cost structure, is a much stronger market force to be reckoned with.

Crown's product mix has shifted from 17 percent solid wood products in 1980 to 9 percent in 1983, as wood products markets declined. Similarly, as newsprint capacity began to exceed demand, Crown reduced its newsprint facilities from 20 percent of total company paper capacity in 1982 to a projected 10 percent in 1985. Growth paper product lines such as tissue and printing and writing papers were expected to grow from 50 percent to 90 percent of company production during the same period.

In 1983 Crown Zellerbach was fifth in US capacity of coated groundwood, with 9.3 percent of total capacity in a segment where the top ten producers own 94 percent. Crown Zellerbach was also seventh in tissue capacity, with 5.9 percent of US total capacity. Crown, traditionally a leader in tissue, cosmetic, and personal products on the West Coast, improved its market position by its sophisticated and ambitious distribution business, which accounted for 31 percent of Crown's sales in 1983. Crown Zellerbach was designated the official paper distributor for the 1984 Olympics.

Georgia-Pacific

Georgia-Pacific Corporation is a major producer of forest products, including lumber, plywood, building products, pulp, paper, and chemicals. In 1982, Georgia-Pacific was the largest forest products firm in the world in terms of gross sales.

Full vertical integration is a key concept for Georgia-Pacific. Thirty-one percent of its sales were in paper, while its chemicals division, 11 percent of gross sales, largely produces chemicals for use in the company's own operations. In 1983 Georgia-Pacific owned or controlled 5.1 million acres of timberlands, 65 percent of which were in the South. This resource yielded 50 percent, by industry estimates, of Georgia-Pacific's total fiber needs.

Sales in 1983 were $6,500 million, an all-time high for the company, with an operating margin of 11.5 percent. Georgia-Pacific has undergone a restructuring to focus on its core businesses: building materials, pulp and paper, and chemicals. Focus has been narrowing on selected paper and fiber markets, especially market pulp, tissue, printing papers, and industrial specialty papers. This narrowing of focus, combined with cost management, was expected to improve the company's competitiveness in domestic and world markets.

In 1983 Georgia-Pacific was second only to Weyerhaeuser in paper grade market pulp capacity, with 6 percent of the North American total, and was the largest exporter of market pulp in the US. Georgia-Pacific was also strong in tissue (9.6 percent of US total capacity) and in uncoated printing and writing papers (5.1 percent of total US capacity).

Georgia-Pacific has withdrawn from its timber joint venture in Indonesia. It still retains manufacturing operations and timber holdings overseas.

International Paper

International Paper is the largest single paper manufacturer in the world, with 1983 sales of $4,357 million, 84 percent of which were generated in pulp, paper, and converted products. International Paper produces a wide variety of types of paper and paperboard as well as solid wood products, market pulp, and converted paper products.

In 1983 International Paper owned or controlled seven million acres of timberland, which supplied 35 percent of the company's total fiber needs. However, size alone did not protect the company from the 1980-1982 recession. Profits declined under the pressure of lowered paper prices and lower operating rates maintained during the general economic downturn. International Paper's response was to shift to more profitable product lines by converting existing machines from basic commodity production to higher value-added items. Through this tactic of conversion rather than purchase of new equipment, International Paper was able to face rising prices in 1983 with a much lower debt load (long term debt equaled 21 percent of capital) and lower overhead costs than many competitors.

Another tactic the company used to reduce the pain of recession was to do some selective business unit pruning, as seen in the sale of Canadian International Paper in 1981 and the divestiture of some forestlands in 1983. Twenty percent of the company's sales still come from foreign operations, however. In 1983 International Paper had foreign operations in five countries other than the US. Overseas activities consisted of pulp and paper mills and corrugated box plants in Western Europe and in Latin America.

In 1983 International Paper's sales mix was divided into the following groups: pulp and paper manufacturing, 31 percent; converted paper products, 44 percent; and wood products, 24 percent. International Paper was number one in US capacity in bleached paperboard (23.5 percent of the total US capacity), coated groundwood and publishing paper (14.9 percent of US total), and linerboard (10.1 percent of the US total). The company also was strong in capacity in corrugating medium (third in the US, with 7.4 percent of total) and in uncoated printing and writing papers (again third, with 7.5 percent of US total capacity). International Paper was also a significant shipper of market pulp.

James River

James River is a remarkable company. In 1982 it moved from the 65th to the 15th ranked paper company in the world in gross sales. This made James River the largest independent producer of specialty papers in the US. The huge expansion was largely a result of a very aggressive acquisition and merger policy. In 1983, James River owned five pulp mills, 22 paper mills with 60 machines, and 32 converted products manufacturing plants.

James River owns or has cutting rights on 3 million acres of timberland, making the company 70 percent internally fiber sufficient. Manufacturing facilities serve widely diversified special, industrial, consumer, and technical markets, and the company produces market pulp and commodity grades of paper as well.

Sales in 1983 were estimated at $2,200 million. Sizable recent acquisitions in food service and packaging products added stability to the mix, as well as contributing 65 percent of sales and most of the profits for the company in FY 1982. Operating margins in 1983 were estimated at 11 percent and rising.

The firm's long-term debt to capital ratio was 47 percent in 1983, highly leveraged in an industry where the typical ratio is 30-35 percent. This figure has ranged as high as 68 percent in the previous five years for James River, however, and seems likely to remain high as long as the acquisition rate is so high.

In 1983 all of James River's operations were domestic and tended to be focused on specialty markets where competition was low, margins were relatively high, and quality and service factors outweighed economies of large scale so important in commodity paper markets. Most of James River's acquisitions were small, established paper companies with problems competing with big, well-financed commodity producers. With consolidation of these mills into a larger operation with superior marketing and product development, the possibilities seemed brighter.

In 1983 James River owned 4.9 percent of US bleached paperboard capacity and 8.6 percent of US tissue capacity (the company was the fifth largest producer in the country in tissue).

Kimberly-Clark

Kimberly-Clark is a major world producer of consumer paper products, second only to Scott Paper. Kimberly-Clark is less integrated than most American forest products companies, with only a small percentage of its fiber supplied from owned sources. Kimberly-Clark's specialty is the production, development, and marketing of all types of consumer personal paper products.

Sales in 1983 were $3,274 million, broken down by the following product lines:

Consumer products	69 percent
Newsprint/market pulp	16 percent
Paper products	13 percent
Other	2 percent

Since consumer paper product demands are relatively inelastic with respect to recession, inflation, and business cycles, Kimberly-Clark's earnings have not varied much compared with those of other forest products firms whose wood products segments, for example, are sensitive to interest rate changes. Operating margins for Kimberly-Clark for (1980-83) were stable at about 12.5 percent, despite a fairly severe recession.

In 1983 Kimberly-Clark had manufacturing operations in 22 countries, as well as an active overseas marketing program. At home, Kimberly-Clark has been industry leader in newsprint capacity (ninth in North American capacity, with 3.8 percent of total) and in tissue capacity (sixth in the US with 7.5 percent of its total). Kimberly-Clark's R&D/sales ratios has been higher than most paper companies' due to the consumer-oriented nature of its markets. The company's high level of expertise in absorbency and fiber chemistry is another important, if less tangible, asset.

Louisiana-Pacific

Louisiana-Pacific was formed in 1973 as a spinoff of 20 percent of the assets of Georgia-Pacific. It shares with James River the distinction of being the only new entrants into the US pulp and paper industry in recent years. The comparison is especially interesting in that the orientations of the two companies could not be more different. While James River is a "paper company's paper company" and specializes in product differentiation, Louisiana-Pacific takes the opposite position and generally will not integrate much further forward than the production of market pulp, although the company does produce linerboard and shipping containers. Additionally, the company is less fiber sufficient than many US companies, preferring to buy logs on the market than to hold capital-intensive timberlands.

This nonintegrative policy decision explains why Louisiana-Pacific's product mix in 1983 was only 20 percent paper (12 percent from pulp, 9 percent from linerboard and containers) and the remainder from solid products. Louisiana-Pacific is currently the third largest producer of lumber in the US and is fast becoming the market leader in waferboard construction. Involvement in the fiber side seems to be declining, in fact, with the sale of the company's pulp facilities in Ketchikan, Alaska, to employees.

This strategy, contrary to industry convention, appears to be paying off. Louisiana-Pacific's 1983 sales of $1,102 million showed a more than 25 percent increase over 1982's performance. The nonintegrative policy and extremely tight cost controls have kept cash flows positive and profits up. Until 1979, operating margins hovered around 20 percent, which is high for the industry. The results for 1983 of 14 percent reflected a bitter six-month-long labor strike which was in effect at 18 of Louisiana-Pacific's western sawmills, although shipments were nearly back to prestrike levels, using nonunion labor.

The company's 1983 debt was a very low 19 percent of capital, in line with management's desire to stay liquid in order to take advantage of new opportunities. Louisiana-Pacific also carried high levels of liquid assets for the same reason.

Scott Paper Company

Scott Paper Company is the world's largest producer of consumer paper products, with 1983 sales of $2,450 million. Scott owns 29 plants and mills in the US, with manufacturing operations in 17 foreign countries. With 92 percent of its 1983 revenues coming from pulp and paper and converting, Scott generated 43 percent of the fiber it needed internally, using wood from the 3.3 million acres of timberland the company owned or controlled. Recent attempts to sell large blocks of timberland in the Pacific Northwest, if successful, could reduce that internal fiber sufficiency.

With 65 percent of its 1983 sales in converted paper products, Scott was unusually recession resistant, with its 1983 operating margin of 11.5 percent near the company average for the previous five years. Scott is now at the end of a very ambitious five-year, $1.7 billion capital spending program to raise productivity, lower costs, and improve product mix profitability by shifting to faster-growth lines. Marginal businesses have been disposed of to improve overall return on owners' equity including some marginal mills in the Pacific Northwest.

All this restructuring should make Scott an even more formidable competitor in world markets.

St. Regis Corporation

St. Regis Corporation is a fully integrated forest products company and a major producer of kraft paper and board, printing papers, packaging, and wood products. The 5.9 million acres of timberland owned or controlled by St. Regis in 1983 yielded 44 percent of the solid wood and 35 percent of the pulpwood required by the company's operations.

A rough breakdown of St. Regis's 1983 sales of $2,775 million shows:

Paper production	54 percent
Packaging and converting	30 percent
Construction products	15 percent
Energy	1 percent

From 1980 - 1983, St. Regis disposed of a dozen marginal mills, as well as cut its work force by 10 percent, in an attempt to improve its position coming out of the recession. The operating margin in 1983, estimated at 8 percent, was low compared to the company's pre-1980 average of over 13 percent.

With 132 plants in seven countries (12 percent of sales come from foreign operations), in 1983 St. Regis was a major competitor in domestic and foreign markets. The average St. Regis plant, estimated to be 12 years old, is somewhat older than the industry average. The company is also saddled with significant newsprint capacity, likely to be less than competitive in view of projected North American, and possibly world, overcapacity in that commodity.

In a somewhat brighter market, in 1983 St. Regis owned 14 percent of the coated groundwood capacity in North America, which is a very concentrated segment, with 94 percent of all capacity in the hands of the top ten producers. Coated groundwood is expected to grow as direct sales in newspaper supplements and direct mail catalogs expand. St. Regis also is the fifth largest US producer of recycled paperboard, a product most traditional papermakers have ignored. In 1983 the company owned 6 percent of US uncoated groundwood printing paper capacity as well.

St. Regis holds strong positions in packaging, kraft, and linerboard.

Weyerhaeuser Corporation

Weyerhaeuser, with six million acres of prime timberland in the Pacific Northwest and the South, and controlling over ten million more in Canada and Southeast Asia, supplies all its own timber needs, which makes it unique among its large competitors. Its 1983 sales of $4,883 million were divided roughly as follows:

Building materials (lumber and panels)	53 percent
Pulp and paper	25 percent
Containers and packaging	14 percent
Other (mortgages, real estate)	9 percent

While Weyerhaeuser has long been a dominant presence in the domestic and export lumber and log business, a $2 billion capital expansion program during the early 1980's made Weyerhaeuser's paper operation a top world class competitor in its own right. With 20 percent of the total company sales generated by foreign operations in five countries, mostly in the Pacific Rim, Weyerhaeuser dominates the region's forest products trade.

In 1983 Weyerhaeuser was seventh in US bleached paperboard, with 4.9 percent of the total capacity. The company had even more market presence in linerboard and corrugating medium, with 8.7 and 7.1 percent of the capacity in those commodities. This made Weyerhaeuser the number two US producer of linerboard and number three in corrugating medium. The company also owned 4.9 percent of the nation's capacity in fast-growing uncoated printing and writing papers. Not surprisingly, given its large timber base, Weyerhaeuser was also the top North American producer of paper grade market pulp, with 8.5 percent of total capacity in North America.

With these highly competitive capacities in place (average plant age: six years), Weyerhaeuser seems certain to become even more firmly entrenched as a market leader in the US and the Pacific Rim.

Operating margins, roughly 15 percent in 1983, were recovering from 1982 recession lows of 12.7 percent, a figure many companies would find acceptable in any year. Weyerhaeuser's huge timber base has subsidized its growth and surely represents a considerable portion of its earnings performance. Given a policy of modest long-term debt, a strong positive cash flow, and capital outlays largely completed, Weyerhaeuser is now in an excellent position to improve its Pacific Rim market dominance in the future.

Willamette Industries

Willamette Industries of Portland, Oregon, is a rather closely held forest products firm, 23 percent owned by directors and their families, many of whom are related to original owners. The company was formed by the merger of six related forest products companies in Oregon in 1967.

In 1983 Willamette owned 220,000 acres of timberland in Oregon and an additional 336,000 acres in Texas, Louisiana, Arkansas, and Tennessee, with purchase contracts on 150,000 more. These resources provided over 50 percent of the wood required to operate Willamette's 60 diversified timber, panel products, and paper and paperboard facilities.

The company is highly integrated forward as well; the wood products plants provide chips for the paper mills, and most of the company's paper production goes to company-owned converting plants to produce shopping bags and shipping containers of all varieties. Willamette has also diversified into business forms, the fastest-growing segment of the US paper market, which includes computer papers. The company took a cautious approach to this product at first, building the forms conversion plants and establishing a market before integrating backward to the production of fine paper with a new machine completed in 1981. Since the market was already in place, there were no difficulties finding customers to absorb the 100,000 tons of capacity from the new machine. The company's capacity in this product doubled from 1978 to 1983.

Willamette's product mix has changed significantly over the years, from about 50-50 wood products and paper products in 1977 to a split more like 35-65 in 1983. Its total sales in 1983 totaled $1,046 million, and although its size is modest, its returns to shareholders are among the best in the industry. Canny management and lean structure are generally credited with saving Willamette from faring as badly in 1980-82 as most of its competitors. The salaried personnel layoffs so common to virtually all companies in the industry were not seen at Willamette, and the company was able to maintain its $1.50 per share dividends without a break. *Fortune* magazine calculated that total average returns to shareholders including dividends for the ten years ending in 1982 averaged 11.98 percent, which placed Willamette well above many industry giants, including International Paper, Champion International, and Weyerhaeuser. Indicative of this high return is the fact that Willamette had a higher physical output per share of forest products than any other company.

Appendix 2

CANADIAN COMPANY PROFILES

Abitibi-Price

In 1983 Abitibi-Price was the largest producer of newsprint and of uncoated groundwood paper in North America, with 12.5 and 11.4 percent respectively, of total North American capacity. Although all of its newsprint, groundwood, fine paper, and kraft mills were located in the Eastern provinces of Ontario, Quebec, and Newfoundland, Abitibi-Price also owned a nationwide system of converting (mostly envelopes and stationery items) and distribution operations. Fiber is mostly supplied from over 11 million hectares of forestland which Abitibi-Price owns or controls, under license from the government. Abitibi-Price produces little market pulp; most of its production is consumed internally, and most of its paperboard goes to Canadian box manufacturers.

Total sales for Abitibi-Price in 1982 were $1,632 million (Canadian) reflecting up-and-down markets since 1978, when sales were $1,292 million. Fifty-four percent of the company's output went to US markets, mostly in newsprint, with 34 percent going to Canadian markets and 12 percent to other markets, mostly Europe. The geographic market share has varied little over time.

Abitibi-Price's 1983 product mix consisted of the following:

Newsprint	49 percent
Groundwood	13 percent
Fine papers*	23 percent
Kraft materials	4 percent
All pulp and paper	89 percent
Lumber and building products	11 percent

*Includes converting and distribution.

Additions to fixed assets from 1979 to 1982 averaged more than $185 million per year, compared with less than 17 million per year between 1970 and 1974. Most of this accelerated capital expenditure went to modernization of existing capacity, including making machines faster and more efficient, converting to more energy- and fiber-economical processes, and generally cutting costs. The company is now a leader in uncoated groundwood and is converting some newsprint capacity to this fast-growing, more profitable grade.

Abitibi-Price also engaged in some internal rationalization by pruning unprofitable and inefficient businesses, mills, and machines. Profitability declined for Abitibi-Price as for most Canadian paper companies in the past several years:

1978	1978	1980	1981	1982	
7.9%	10.2%	7.7%	7.4%	3.6%	Return on sales (after tax)
23.1%	27.1%	16.3%	17.4%	7.4%	Average return on shareholders' equity

The direct link between operating rate and profitability can clearly be seen when one compares 1979, during which Abitibi-Price operated at 100 percent of capacity, and 1982, when its newspaper operations were running near 80 percent of capacity.

Abitibi-Price became caught in the cost-price squeeze common in Canada, when costs rose faster than US costs due to differential inflation rates and US producers setting market prices. Frequent and prolonged labor strikes also negatively affect Abitibi-Price's profitability. Favorable factors included ownership of several hydroelectric dams which reduced power costs and the US-Canadian exchange contribution. The latter factor was significant, contributing $16.6 million (46 percent) after-tax net earnings in 1977. It was calculated in 1976 that each one-cent change in the US-Canadian exchange rate affected earnings per common share by 11 cents.

Abitibi, after a long equity ownership in Price, acquired 100 percent of Price's outstanding shares in 1978 to form Abitibi-Price. The company had no overseas operations, although it had hardboard manufacturing facilities in the US in Mississippi, Ohio, North Carolina, and Indiana. Additionally, Abitibi-Price in 1968 acquired 50 percent of Boise-Price Southern Newsprint at DeRidder, Louisiana, and became a 50 percent joint venture partner with Thomson Newspapers in a two-machine newsprint facility in Augusta, Georgia.

British Columbia Forest Products

One of Canada's largest integrated forest products companies, British Columbia Forest Products (BCFP) was incorporated in 1946 and as such is a relative newcomer to the industry. All of its manufacturing operations are in British Columbia, on Vancouver Island and in the interior. BCFP also owns part or full control of several related companies:

Company	Percent owned	Location	Type of Operation
Blandin Paper & Forest Products	100%	Minnesota, US	Coated stock Aspen waferboard
Swiftsure Towing Co.	100%	Vancouver, BC	Coastal tug & barge
Croftech Inc.	100%	Vancouver, BC	Engineering project management
Donohue-St. Felicien	40%	Quebec	Woodland, saw & pulp
Finlay Forest Products	43%	Quebec	Pulp and sawmills
Pinette & Therien Mills	50%	Quebec	Lumber mills, timber
Western Forest Products Ltd.	33%	Quebec	Two pulp mills

Like many Canadian companies, BCFP is fully back integrated to the forestland, but is not integrated forward much past the commodity paper stage. Total sales for 1982 were $647.6 million (Canadian), 70 percent of which was generated by paper and pulp sales, and 84 percent of which came from export sales, mostly to the US.

Sales by markets were:

US	Canada	Europe	UK	Japan	Other
57%	16%	8%	5%	3%	11%

and sales by product:

Paper	48%	Lumber	18%
Pulp	22%	Plywood/waferboard	6%
		Other building	6%
Pulp and paper	70%	Total lumber & waferboard	30%

For Canadian lumber producers 1982 was a poor year due to continued poor construction conditions across the border; while BCFP was building and starting up new sawmills in British Columbia and Alberta, it was also closing and curtailing others. Conditions in pulp and paper were not much better. Pulp prices began 1982 at $565-580 (US) in the United States. This was not good news for a company which owned a significant amount (4.2 percent) of the North America paper-grade pulp capacity. In spite of newsprint overcapacity, BCFP also brought on line in 1982 a new newsprint machine and remodelled and speeded up another. As a result, BCFP had the dubious honor of being the largest money loser in the North American industry in 1982 ($49.4 million).

Mead Corporation, a large US paper manufacturer, holds some (unspecified) interest in BCFP. Mead was the largest paper industry money loser in 1982 in North America with $85.7 million in losses (US).

Consolidated-Bathurst

Consolidated-Bathurst is a large Canadian forest products firm fully integrated forward from the woods (where it obtains about half its fiber needs from company limits and freeholds), to packaging and converting. The company produces kraft pulp and linerboard, corrugating medium, kraft paper, boxboard, and newsprint.

Consolidated-Bathurst was the first North American newsprint manufacturer to produce in Europe. It bought Bowater's newsprint mill in

Ellesmere Port in 1981 for about $15 million and planned improvements of $27 million.

Total sales in 1982 were $1,247 million (Canadian), 78 percent of which came from pulp and paper sales.

Product mix breakdown (1977):

Newsprint	27%	Packaging (domestic)	12%
Pulp	7%	Domglas	20%
Paper, container, lumber	12%	Overseas packaging	22%
Pulp and paper	46%	Packaging	54%

Domtar

Domtar is a highly integrated Canadian forest products firm which produces and markets fine papers, newsprint, groundwood specialties, pulp, kraft paper and linerboard, boxboard, containers, and container board. It is Domtar's avowed strategy to concentrate on businesses in which the company can maintain a major market share and can be competitive with the lowest-cost producers and in which growth possibilities are attractive. Vertical and horizontal synergy are important elements.

In 1983 Domtar owned 875,000 acres of timberland, with cutting rights to ten million more, supplying 50-60 percent of the company's fiber needs. The company also produces sufficient kraft pulp for all its internal needs, selling surplus as market pulp. Virtually all company facilities are located in Eastern Canada except for a nationwide scattering of corrugated box plants and wholesale distribution facilities in the UK and Western Canada.

Total 1982 sales were $1,687 million. Pulp and paper sales, 49 percent of the total, were generated in the following segments:

Fine papers	28%	
Newsprint, groundwood papers	14%	
Other pulp and paper	8%	
Total		49%

while packaging included:

Corrugated container & board	13%	
Other packaging	2%	
Total		15%

Construction materials (20 percent) and chemicals (16 percent) accounted for the rest.

Domtar's fine papers segment includes the two largest fine paper mills in Canada, which allowed Domtar to capture the largest single share of the Canadian fine papers market. Domtar's 15 corrugated container facilities span all Canadian regional markets. Domtar maintains the largest integrated containerboard facilities in Canada.

In line with Domtar's aggressive business strategies for cost competitiveness, the company spent $676 million (average $169 million per year) from 1979 to 1982 on maintenance, improvement, and expansion of its businesses, nearly doubling its net fixed asset base ($402 million in 1978 to $779 million in 1982). Of this total, 31 percent went into the pulp and paper group while 17 percent went into the packaging group, all for cost reduction, pruning marginal operations, and "streamlining."

One set of costs which won't be reduced by more investment is the recurring cost to Domtar of annual labor strikes. In 1982 the company negotiated 77 collective labor agreements, not all of them initially successfully. Domtar has a history of recurrent costly labor strikes in all its divisions. Strike-caused shutdowns lose customers and market shares, as well as more directly affecting profits.

Another concern for Domtar, the largest supplier of fine papers in Canada, is the phasing out of the existing protective tariffs on fine paper exports to Canada. Scheduled to be reduced 50 percent by 1987 in line with a GATT agreement, these may make it more difficult for Domtar to continue to be the low-cost producer.

Domtar's profitability over time:

	1982	1981	1980	1979	1978
Net After Tax Income/Sales	[na]	3.6%	5.7%	6.5%	5.1%
Return on Sales/ Net Equity	1.4%	9.8%	18.6%	23.3%	17.5%

Canadian exchange rates against the US dollar were a source of improved performance at Domtar. A one-cent change in US/Canadian exchange rates was estimated to have a $1.5 million effect on net income after taxes.

Although exports are important to Domtar, the company had no manufacturing facilities outside Canada in 1983.

MacMillan-Bloedel

MacMillan-Bloedel is one of the oldest forest products companies in North America. Originally a product of the partnership between MacMillan Export Company Limited and Bloedel, Stewart and Welch, Ltd., both established in the early 1900's, the company was consolidated in 1960 as MacMillan-Bloedel, Ltd., Inc. The broadly integrated company owns or manages over 4 million acres of timber in British Columbia and Alabama, with lumber, plywood, wafer and particle board, newsprint, groundwood, pulp, paper and board, and corrugated plants in all parts of Canada and the US. Corrugated container plants are located in the US, Canada, and the UK as well. Additionally, MacMillan-Bloedel owns part or all of several trading, wholesaling, and distribution firms in all parts of the world, including the UK, Australia, Hong Kong, Tokyo, Singapore, Malaysia, the Philippines, Antwerp, Paris, and the US. The Hong Kong sales company MacMillan-Jardine is controlled by MacMillan-Bloedel, and this joint venture is a part owner of several other investments around the Pacific Rim.

MacMillan-Bloedel's total 1982 sales were $1,843 million, and its earnings a negative $57.3 million, which made MacMillan-Bloedel the fourth largest money loser in North America in 1982. Seventy percent of the company's newsprint and 54 percent of its total pulp and paper output are exported to the US. Other 1980 markets are shown below:

	Newsprint	All pulp & paper
British Columbia	5%	6%
Other Canada	5%	4%
United States	70%	54%
Japan & Orient	5%	11%
United Kingdom & Europe	1%	14%
Other	14%	10%

MacMillan-Bloedel is more heavily oriented toward lumber and wood products than some of the other Canadian firms, and its location on the West Coast rather than near the large population centers of the Eastern seaboard may be one reason. The product mix for MacMillan-Bloedel for 1980:

Sales

Lumber and panels	Fine paper and newspaper	Pulp	Container and Packaging	Other
40%	25%	12%	21%	2%

Until 1980, MacMillan-Bloedel's strategy seemed to be to diversify in every geographic direction and to market aggressively in every large available market. In 1971, MacMillan-Bloedel had various operations in the US, UK, Hong Kong, Malaysia, Singapore, Indonesia, Australia, Holland, and Spain, and these operations ranged from sales to forestry and logging. MacMillan-Jardine formed other joint ventures for the purposes of marketing, logging, operating wood products plants, and real estate development.

In 1980, MacMillan-Bloedel was in the middle of a strong five-year capital expenditure plan to increase capacity and improve cost containment and productivity when the three-year forest products recession began to make itself felt.

Strongly dependent on US housing as a market for its lumber, MacMillan-Bloedel was hit very hard by the recession, labor strikes (estimated to have cost $42 million in 1981 earnings), and differential Canadian inflation. The company in 1982 completed a decentralization and restructuring designed to reduce costs, cut out unprofitable operations, and improve competitiveness in world markets. Elements of this plan included salary cuts, staff reductions, changes in product mix, and a greater emphasis on high value-added products. At the same time, divestitures and financial restructuring took place to assure continued viability for the company, culminating in the sale of MacMillan-Bloedel's corporate headquarters building in Vancouver, BC. The $62 million received for the building was used to reduce debt.

MacMillan-Bloedel had the third largest newsprint capacity in North America with 7.7 percent of the total, as well as 7.3 percent of uncoated groundwood capacity and a significant share of market pulp capacity. Although strikes are a costly fact of life, MacMillan-Bloedel's costs were somewhat alleviated by ownership of several hydroelectric dams in BC and Ontario.

Profitability for MacMillan-Bloedel, never impressive, slid severely after 1980:

1978	1979	1980	1981	1982
11.6%	16.0%	12.0%	0.1%	(5.7%)

Operating rates in pulp and paper in 1982 were between 70 and 55 percent of capacity; even so, pulp and paper was the only operating segment with positive earnings in 1982. Debt/equity, at 45 percent, was well above the company's target rate of 35 percent.

MacMillan-Bloedel has acquired full ownership of Embrasca, a Brazilian former joint venture with Brascan. On the divestiture side, the company wrote off its 43.8 percent interest in La Cellulose d'Aquitaine in 1979. MacMillan-Bloedel is still active in Mississippi and Alabama with forestland, lumber, plywood, and paper operations. MacMillan-Bloedel, through its MacMillan-Jardine Sales Company, has had good success in Far Eastern markets, notably its successful bid for the highly competitive JTT (Japan Telephone and Telegraph) directory paper in 1983. Traditionally supplied only by Daishowa and Jujo, this contract had never gone to a foreign firm before.

Appendix 3

JAPANESE COMPANY PROFILES

Daishowa Paper Manufacturing

Daishowa paper was the fifth largest pulp and paper company in Japan in 1982, and the largest independent one (all the other large companies being allied with one or more of the great business combines). Daishowa's total 1982 sales were equivalent to about one billion US dollars, with the following product mix:

Printing papers	43%
Newsprint	21%
Containerboard	16%
Whiteboard	11%
Wrapping papers	9%

In 1983 Daishowa consolidated several of its businesses in which it held controlling interests: Daishowa Pulp Manufacturing Company, 60 percent ownership; Daishowa Paper Trading Company, 100 percent ownership; and Daishowa International (overseas development), 91 percent ownership. Daishowa's financial position became rather precarious in the severe recession in pulp and paper in 1981-82, until the company's main bank, Sumitomo, loaned the ailing company some of its own executives to "increase cooperation with the rest of the industry, improve its financial position, and, in general, modernize its management." The recent merger was part of this "progressive normalization," an attempt to reorganize more than 50 subsidiaries and improve efficiency. After the consolidation, Daishowa became the second largest "seaside, pulp-to-paper integrated company" in Japan, second only to Oji Seishi.

In addition to internal reorganization, Daishowa worked to upgrade its product mix to more profitable lines by modifying some of its mills to produce printing papers instead of lower-value kraft paper. A continuing effort to reduce costs and dependence on overseas suppliers of chips has characterized Daishowa's strategy, as well as that of most Japanese companies.

Daishowa, with most other Japanese paper manufacturers, carried a heavy debt burden (income in 1982 covered only 66 percent of interest charges) due to long-range unprofitability and worsening cost considerations. For example, manufacturing costs rose from 88 percent of sales in 1978 to 96 percent in 1981. This was likely due to low operating rates and increases in raw material costs, because cost controls and possibly the structural recession cartels brought the figure back to a more normal 81 percent in 1982. The largest rise for a single component in that time was materials, which rose from 61 percent to 72 percent of sales from 1978 to 1981, and dropped back to 60 percent in 1982. It is not clear whether the improvement may have been due to stronger management from the Sumitomo group, or whether the recession cartels did improve price stability, but probably there were effects from both.

In 1982, Daishowa had a 3 percent share of the total pulp production in Japan and a 9.5 percent market share of the total Japanese paper market, as well as being the third largest paperboard manufacturer in Japan. Daishowa buys virtually all of its raw materials; company forestland ownership is not common as it is in North America.

The company has been active for some time seeking new sources of captive supply for fiber, and this search was intensified after the chipshock of 1979-80. Several ventures have been actively explored, including forestland development and trial plantations, either alone or in consortium with other companies in Malaysia, Indonesia, and Southeast Asia. Several joint ventures and overseas companies were formed, including Daishowa America (US) and Harris Daishowa (Australia), to ship pulp chips to Japan; Cariboo Pulp and Paper in Canada with Marubeni, to produce and export to Japan bleached kraft pulp; and Quesnel River Pulp Company, Canada, to produce and export to Japan thermo-mechanical pulps. The five enterprises together contribute 236,000 tons of pulp and over a million cubic meters of chips per year to Daishowa's raw material supply. Although these operations and agreements were created to allow Daishowa some freedom from the capricious pulp and chip open markets, a catch-22 has now occurred in which lower paper demand in Japan plus greater use of wastepaper as furnish has somewhat reduced the industry's need for imported fiber.

Honshu Paper

Honshu Paper is a loosely connected member of the industry-dominant Mitsui group with ties to the Daiichi Kangyo group as well. It is the largest paper company in Japan in terms of 1982 sales. Relatively integrated by Japanese standards, Honshu has businesses in industrial chemicals, lumber manufacture and sale, and plastics as well as pulp and paper operations. The company is also a major converter, producing corrugated board, corrugated containers, and specialty paper products.

Honshu began as a paper company and split off from the Mitsui Zaibatsu in 1949. The company diversified in container board in 1959, to paper and paperboard converting in 1964, and started producing nonwoven fabrics in 1969. In 1975 Honshu began production of plastic films and packaging, a big growth market in Japan today.

Although Honshu has been moving toward more diversification and higher value-added products, as of 1982 its total sales of approximately $970 million broke out in the following way:

Paper and board	57%
Converted products	31%
Lumber	10%
Chemicals & other	3%

Honshu is regarded as the top producer in the paperboard sector, but linerboard and similar products which require large volumes of virgin fiber are coming under increasing pressure from North American imports. Resulting linerboard overcapacity and depressed prices have made paperboard a less desirable segment of the industry.

As a result, Honshu merged in April 1983 with Fukuoka Paper (producers of linerboard) and Tonshin Paper (corrugated box producers) as well as Kanzaki Paper in an attempt to consolidate its top paperboard position. The company is now integrated from pulp to liner to corrugated box production. Honshu is also the parent company of Honshu Container, which makes corrugating medium.

Not only was Honshu the largest paperboard producer in Japan in 1982, but it also produced 4 percent of the nation's pulp as well. Its manufacturing costs were comprised as follows:

78%	Raw materials
9%	Wages
13%	Other expenses

Debt coverage was only 1.09 times earnings; that is, 1982 earnings exceeded interest payments by only 9 percent. Like most large paper companies in Japan, Honshu's debts are large. One of its avowed strategies is to make an effort to reduce total debt but so far the trend is in the opposite direction. Other strategies involve cost reductions (increase use of wastepaper, cut energy use, reduce personnel levels) and expansion toward the high value-added of converted lines. The year 1982 showed a drop in sales due to the severe downturn in lumber markets, but an overall increase in profits due to cost savings.

In 1982 Honshu exported 8 percent of its total sales. Its major exports were kraftliner, corrugated board, folding boxboard, thin papers, and specialties like condenser paper. Its markets for these items included Latin America, Oceania, Africa, Hong Kong, and even the US and West Germany. Honshu's paper imports are primarily commodities from the US and Finland.

The company is a joint venture partner of Mitsubishi and an unnamed Canadian interests in a group called Crestbrook Forest Industries formed in British Columbia in 1967. The venture ships softwood bleached kraft pulp and lumber to Japan. Finally, in 1971 Honshu joined in a venture called Jant Plywood Ltd. in Papua, New Guinea. The purpose was to develop forests and raw material and chip supplies. The majority of these ventures have been largely abandoned due to restrictive trade legislation and disappointing returns.

Jujo Paper

Jujo Paper, the second largest Japanese pulp and paper company in 1982, also began as part of the Mitsui group, along with Honshu Paper and Oji Paper. Jujo is large by any standard, being the 70th largest company in Japan in total sales. Its product lines include pulp, paper, paperboard, chemicals, lumber, building materials, and landscaping. Its paper business is also very diversified including NCR (carbonless) papers, thermal printing papers, asbestos paper, and cigarette paper as well as the more commodity-type products such as newsprint, printing, and kraft wrapping and packaging papers.

Sales in 1982, approximately $1.3 billion, were distributed as follows:

Printing papers	39%
Newsprint	32%
Converted papers	10%
Converted products	10%
Pulp (market)	7%
Timber and other	2%

The reorganization and rationalization movement in the Japanese pulp and paper industry was reflected in the company's merger with Senju Paper in 1982 to make Jujo Japan's third largest paperboard company.

In 1981 Jujo produced 10 percent of Japan's pulp and held the dominant market share in milk cartons, thermal paper, and NCR papers.

Jujo's overall cost structure in 1982 was as follows:

Raw materials	67%	total	cost
Wages	15%	"	"
Other	18%	"	"

Like most Japanese pulp and paper companies, its interest coverage was minimal: Earnings were 1.02 times interest expense.

Effects of the severe paper recession included an emphasis on cost reduction both by process cost controls and by capital outlay for plant modernization of kraft facilities and new product development. The company also spent about $42 billion on cost reductions and energy savings alone. Other activities included improving quality control, upgrading the marketing system, and consolidating Jujo's 38 subsidiaries.

In 1982 Jujo exported 9 percent of its total sales and had area trading offices in Seattle, Washington; Vancouver, BC; and Dusseldorf, West Germany. The company is involved in joint ventures with two large American firms: Kimberly Clark (Jujo-Kimberly KK, 1963) and Weyerhaeuser (NorPac, 1981). Jujo-Kimberly KK operates mostly within Japan, with two tissue mills which made it the largest tissue manufacturer in Japan in 1979. Norpac operates at the Weyerhaeuser site in Longview, Washington, exporting 160,000 tons of newsprint to Japan in 1982. In response, Jujo's domestic plants had to cut their production to compensate.

Additionally, Jujo is involved in a joint venture with Findlay Forest Industries in British Columbia which supplies 60,000 metric tons of high-grade bleached groundwood pulp to Japan per year, as well as lumber.

Jujo is also involved in M.D.I., a consortium of Japanese pulp and paper companies, to promote development of forest raw materials sources in Sabah and Indonesia.

<u>Oji Paper</u>

Oji Paper Company is the oldest and largest paper company in Japan. Founded in 1873, it was the first company to machine-manufacture paper in Japan, and prior to World War II Oji produced 80 percent of Japan's paper. After the war, in 1949, the company was separated into three major parts which became Jujo Paper Co., Honshu Paper Co., and today's Oji. Mitsui still has ties, particularly through its bank, to all three companies, and the Mitsui Company owns 119,000 hectares of forestland, making it the sixth largest private landholder in Japan. Much of the forestland is on the northern island of Hokkaido.

Oji is composed of 64 affiliated companies, most in the businesses of pulp, paper, and forest products, or in related businesses such as chemicals, forestry, mining, and electric power generation. Some, however, are as far afield as manufacturing and sales of cornstarch, smoked salmon or Chinese liquor.

Oji began manufacturing fine papers in 1952, advanced to kraft and converting capacity in 1962, and after several mergers by 1982 was integrated from commodity papers to high-grade specialties:

Printing & writing papers	38%	
Newsprint	37%	
Tissue & miscellaneous	13%	
Wrapping & packaging	6%	
Total paper		94%
Market pulp	1%	
Converted products	3%	
Logs and timber	2%	
		6%
		100%

Although it is unclear how much of Oji's fiber is internally produced and how much is purchased domestically, the 1982 company bought 36 percent of its pulpwood and 98 percent of its purchased pulp on the export market. In 1980, on the heels of the "chipshock," Oji began to diversify its sources of raw materials, even importing some chips from the US South via the Panama Canal.

In 1982 Oji's raw material imports came from the following sources:

Pulpwood	Pulp
69% US	77% New Zealand
8% Canada	15% Brazil
7% USSR	8% Canada
7% New Zealand	100%
7% Australia	
2% Southeast Asia	
100%	

A strategy of Oji during industry restructuring was to absorb companies like Nippon Pulp Industry, which was merged with Oji in 1979.

This increased and diversified Oji while refraining from adding any capacity to the industry total. Other major company strategies appeared to be the same as seen throughout the Japanese industry--energy conservation, cost reduction, increased machine and employee productivity, and a decreased reliance on imported raw materials. Another strategy was to increase paper exports to Southeast Asia and China to compensate for increased North American imports of paper items to Japan. Oji's exports increased from 2.7 percent of sales in 1979 to4.1 percent in 1982.

Three factors make Oji likely to remain competitive:

1. Greater degree of fiber sufficiency than is common in Japan
2. Some significant energy cost savings from company-owned hydro-dams
3. A deserved reputation for technical development in manufacturing and in new product development.

In 1982, Oji had a commanding Japanese market share in most important major grades of paper:

RANK	PAPER GRADE	PERCENT OF JAPANESE MARKET
1	Newsprint	28.5
1	Packaging and wrapping	15.5
2	Printing and writing	15.0
2	Coated stock	17.6
2	White board	16.9
3	Tissue	5.0
4	Miscellaneous papers	9.1
1	All paper	16.5
1	All paper and paperboard	10.0

Additionally, Oji Seishi was the top pulp producer in Japan, with 14 percent of the market.

Debt has been a problem for Oji, as with most companies in this structurally depressed industry, but for Oji, whose earnings were 1.55 x interest changes in 1982, the problem was somewhat less severe.

Oji is involved in several foreign joint ventures. A joint venture with Canadian International Paper's Dalhousie Mill in New Brunswick, Canada, was designed to supply Oji with 110,000 tons of printing paper per year beginning in 1983. A joint venture pulp mill in New Zealand with Sanyo-Kokusaku Paper and Carter-Holt, Ltd. supplies thermo-mechanical pulp for the partners. In 1979 Oji began working on a technical cooperation agreement with the People's Republic of China involving pulp and paper production, and in 1981 Oji entered an agreement with International Paper for a printing paper mill on the US East Coast, and Oji Paper Company (US) Ltd. in Seattle sells and buys chips, pulp, and sometimes lumber.

Along with these manufacturing joint ventures, there are many ventures to secure raw materials in Sabah, Kalimantan, Brazil, Malaysia, and the Solomon Islands, which Oji has undertaken in conjunction with groups of Japanese paper makers.

Sanyo-Kokusaku

Sanyo-Kokusaku, a member of the Fuyo industrial grouping, is affiliated with the Fuji Bank and with the Marubeni Trading Company. In 1982 it was the third largest pulp and paper company in Japan, as well as being numbered among the top 100 companies in the country overall, in terms of sales. The parent company of Japan Paper Industry Company, Chiyoda Shigyo (producer of kraft products), and Mishima Paper company (tissue), Sanyo-Kokusaku is more highly integrated than most Japanese companies in the industry. The company sales were about $1 billion in 1982, but the total group sales were closer to $1.5 billion. The company's product mix was:

Paper	52%
Pulp	15%
Building materials	11%
Converted paper	4%
Chemicals	5%
Lumber and other	13%

Sanyo-Kokusaku's response to the 1979 "chipshock" was to develop potential resources in Papua-New Guinea, in conjunction with Nissho-Iwai

Trading Company, by establishing a 12,000 hectare plantation. The project, like many others begun about this time, was postponed when demand for chips, and therefore prices, dropped in 1982-83. Also in 1980, Sanyo-Kokusaku signed a ten-year agreement through Maubeni to buy alder chips from Portland, Oregon.

Long a leader in Japanese pulp and paper, Sanyo-Kokusaku's tissue production capacity was the second largest in the country in 1979, and in 1981 Sanyo-Kokusaku had a 9 percent share of the total Japanese pulp production. Of the company's sales, 5.1 percent went for export, and there were several overseas investments, primarily in order to secure fiber supply. The most significant of these are the Sanyo-Scott joint venture (50-50) and the Carter Oji Kokusaku Pan Pacific Ltd. venture based in New Zealand, which ships lumber and refiner groundwood pulp to Japan.